3 8027 00631567
AF838271

SPLICING LIFE?

Cardiff Papers in Qualitative Research

About the Series

The Cardiff School of Social Sciences at Cardiff University is well known for the breadth and quality of its empirical research in various major areas of sociology and social policy. In particular, it enjoys an international reputation for research using qualitative methodology, including qualitative approaches to data collection and analysis.

This series publishes original sociology research that reflects the tradition of qualitative and ethnographic inquiry developed at Cardiff in recent years. The series includes monographs reporting on empirical research, collections of papers reporting research on particular themes and other monographs or edited collections on methodological developments and issues.

Splicing Life?

The New Genetics and Society

PETER GLASNER
ESRC Centre for the Economic and Social Aspects of Genomics, Cardiff University

HARRY ROTHMAN
Institute for Enterprise and Innovation, Nottingham University Business School

ASHGATE

303.483
GLA

© Peter Glasner and Harry Rothman 2004

All rights reserved. No part of this publication may be reproduced, stored in a retrieval system or transmitted in any form or by any means, electronic, mechanical, photocopying, recording or otherwise without the prior permission of the publisher.

Peter Glasner and Harry Rothman have asserted their right under the Copyright, Designs and Patents Act, 1988, to be identified as the authors of this work.

Published by
Ashgate Publishing Limited
Gower House
Croft Road
Aldershot
Hants GU11 3HR
England

Ashgate Publishing Company
Suite 420
101 Cherry Street
Burlington, VT 05401-4405
USA

Ashgate website: http://www.ashgate.com

GLASGOW
CALEDONIAN
UNIVERSITY
LIBRARY

British Library Cataloguing in Publication Data
Glasner, Peter
Splicing life? : the new genetics and society. – (Cardiff papers in qualitative research)
1. Genetic engineering - Social aspects 2. Genetic engineering - Economic aspects 3. Genetic engineering - Political aspects 4. Biotechnology - Social aspects 5. Biotechnology - Economic aspects 6. Biotechnology - Political aspects
I. Title II. Rothman, Harry, 1938-
306.4'6

Library of Congress Cataloging-in-Publication Data
Glasner, Peter E.
Splicing life? : the new genetics and society / Peter Glasner and Harry Rothman.
p. cm. -- (Cardiff papers in qualitative research)
ISBN 0-7546-3238-5
1. Medical genetics--Social aspects. 2. Human genetics--Social aspects. 3. Genetic engineering--Social aspects. I. Rothman Harry, 1938- II. Title. III. Series.

RB155.G56 2004
303.48'3--dc22

2003063798

ISBN 0 7546 3238 5

Printed and bound in Great Britain by MPG Books Ltd, Bodmin, Cornwall.

Contents

List of Tables		*vi*
Acknowledgements		*vii*
Chapter 1	Introduction	1
Chapter 2	The Hunt for the Holy Grail: Compiling the Book of Life	7
Chapter 3	Doing the Human Genome Project	27
Chapter 4	Managing Genetic Information	53
Chapter 5	'Frankenstein' Foods, or the Revenge of the Genetically Modified Potatoes	65
Chapter 6	Globalisation and the Transformation of Nature	77
Chapter 7	From Commodification to Commercialisation	89
Chapter 8	Rights or Rituals: Involving the People	111
Chapter 9	New Genetics, New Millennium, New Society?	121
Bibliography		*129*
Index		*141*

List of Tables

2.1	Origins of the Human Genome Project	10
2.2	Desired Technologies for Sequencing the Human Genome	21
2.3	Timeline of Important HGP Instrumentalities	22
3.1	Timeline Human Genome Project 1991–1997	28
3.2	Timeline Human Genome Project 1998–2003	34
3.3	US Genome Project Expenditure	39
3.4	Wellcome Trust Funding to UK Genome Programme	40
3.5	Global Public Genomic Research Funding	40
3.6	Chromosome Assignments for the Public HGP Centres	43
7.1	Leading US Genomics Companies	97
7.2	Incyte Collaborations	98
7.3	CuraGen Technology Platform	101
7.4	CuraGen Collaborations and Alliances	101
7.5	A Selection of Millennium's Strategic Alliances	104
7.6	Millennium: Mergers and Acquisitions	105
7.7	Affymetrix Collaborations	106
7.8	Celera Genomics Share Price January–December 2000	108

Acknowledgements

This monograph is the result of a programme of work called 'Our Genetic Inheritance', begun at the time we established the Science and Technology Policy Unit as a designated research centre of the University of the West of England, Bristol in 1987. We would like to thank those staff and students at UWE who supported SCITEC and its work in this area until its demise in 2001. In particular we would like to express our appreciation to our various research staff, including Cameron Adams, David Travis, Peter Scott and Wan Ching Yee. Without their hard work on our behalf, much of this book would never have been written.

It would be invidious to single out only a few of our many friends and colleagues who have helped us clarify our thoughts and ideas as the Human Genome Project unfolded. However, Harry Rothman would like to thank Julian Lowe for inviting him to give a series of seminars in the School of Business at the University of Ballarat, Victoria, Australia in 2001, which form the basis of Chapter 7. Peter Glasner would like to thank David Dunkerley who contributed equally to evaluating the case study discussed in Chapter 8. We would also like to thank David Green, originally of Carfax and now at Taylor and Francis, for his faith and support in our successful venture to establish New Genetics and Society as a leading international journal in this field. We have greatly benefited from the inputs of its authors, reviewers and editors. Our thanks go to Sean Rothman for his editorial help and advice. We are also delighted to thank Jackie Swift for her patience and good humour while producing camera ready copy for the publisher.

We would like to thank those organisations who funded some of the research on which this book is based, including the Economic and Social Research Council, the Medical Research Council (Human Genome Mapping Project Resource Centre, Hinxton), the Department of Trade and Industry, The British Council, the Welsh Institute for Health and Social Care, the European Commission (Institute for Prospective Technological Studies, Seville), and the University of the West of England.

Some of the material in this book is developed from papers we have given, and from articles that we have published over the years, including: H Rothman 1994 (Chapter 2); P Glasner 1996 (Chapter 4); P Glasner 2001 (Chapter 5); P Glasner and H Rothman 2001 (Chapter 6); P Glasner and D Dunkerley 1999 (Chapter 8) and P Glasner 2002 (Chapter 9).

Chapter 1
Introduction

The new genetics (including genetic engineering, mapping of human and other species, genetic diagnosis and therapy) forms the scientific basis of new technologies unlike any other, with significant implications for life in the twenty-first century. These geno-technologies directly affect us at a deeply personal level; they pose a threat to the boundaries which conventionally define selfhood and distinguish humans from other animals; they generate potentially novel risks and dangers, with possibly unforeseen, and often unknowable or irreversible outcomes; and they threaten the very basis of accepted understandings of culture and society. They also grow in a complex political, economic and organisational milieu involving science, medicine, commerce and the law in the context of late modern society characterised by risk, reflexivity and globalisation. The new geno-technologies therefore appear to question the very boundaries of Nature itself. Recognising how these boundaries are configured contributes to understanding how society reinvents them through imposing order, reimposing control, and providing space for decision-making and the development and application of policy. It also allows for a critical analysis of the degree to which such processes serve the interests of the many constituencies involved.

The New Genetics

The purpose of this chapter is to introduce the idea of the new genetics, and its historical relation to classical genetics and biotechnology. We outline why the new genetics is considered qualitatively different from other new technologies in terms of its impact. The communist geneticist J.B.S. Haldane in 1946 saw the future of genetics as being bound up with enumerating and locating all human genes and elucidating their biochemical functions; to provide us with 'an anatomy and physiology of the human nucleus ... [giving] the possibility of a scientific eugenics, ... [it] may give us the same powers for good or evil over ourselves as the knowledge of the atomic nucleus has given us over parts of the external world' (cited by Dawes 1952: 181). These were prescient words, but when spoken classical genetics was in no position to deliver on such goals; for that to happen a new genetics would need creating.

Elsewhere we have outlined how this 'new genetics' came into being (Glasner and Rothman 2003). The story in brief is this. Classical genetics developed during the first 50 years of the last century after the rediscovery in 1900 of Gregor Mendel's work on the basic laws of inheritance. The familiar classical concepts and terminology of 'gene', 'mutation', 'phenotype' and 'genotype', and so forth were developed over several decades, along with a body of techniques such as linkage, mapping, and mutagenesis by radiation and chemicals. By the 1930s genetics was flourishing, and influencing and changing evolutionary theory, as well as spinning off sub-disciplines such as medical genetics. By the 1940s it was itself beginning to be influenced by the then new field of molecular biology, which focused attention on understanding the physical basis of life, and especially the gene. By the beginning of the 1950s several researchers were convinced that the answer lay in the structure of DNA. In 1953 James Watson and Francis Crick published their double helix model of DNA, based upon crystallographic work initiated by Rosalind Franklin and Maurice Wilkins; this has been hailed by some as 'the scientific discovery of the century', and is the point at which we can say a synthesis of genetics and molecular biology had arrived.

This discovery stimulated new lines of research producing new concepts and techniques with profound implications for genetics. These discoveries included messenger and transfer RNA, the genetic coding mechanism, replication, transcription and translation. To study such matters new laboratory research techniques had to be invented, amongst which were DNA sequencing, cloning, restriction enzyme, and DNA libraries. These and other methods made it possible to create recombinant DNA through genetic manipulation techniques. This marked not only a scientific turning point but also a technological and commercial one. Here scientifically we may speak of 'the new genetics', which takes classical genetics to a new level, the bio-molecular level. Technologically it allowed massive new possibilities, way beyond the breeding techniques utilising classical genetics. Commercially it opened up a new economic space within which novel products and processes might be created and marketed; biotechnology was the name applied to the industrial activity built on the new genetic technology.

Biotechnology

The term 'biotechnology' has, as Robert Bud (1993) has shown, been used since the early 20th Century to mobilise the creation of technologies from living things. The early goals were mostly related to fermentation-based technology and products, such as Chaim Weizmann's process for acetone production; followed in the post Second World War era by the great success of fermentation of antibiotics, which laid the foundation for today's pharmaceutical industry. However, contemporary understanding of biotechnology has now come to be associated with the great scientific advances in recombinant DNA in the 1960s and 1970s. These encouraged the development of various techniques for manipulating genetic materials produced from certain species in the laboratory and transferring them to different organisms, usually bacteria, within which the transferred gene could be propagated. This technique, popularly known as genetic engineering, became the basis for a new biotechnology 'industry' spawning a host of new companies, particularly in the US, funded by excited venture capital and stock markets. The excitement impacted on politicians, and there were in the early 1980s throughout the industrialised world many government policy reports speculating on how biotechnology might be harnessed to industrial growth. These developments were not universally welcomed, some people fearing that the new technology might release novel and dangerous organisms into society and the environment. Consequently there were vigorous debates in some countries, notably the US and UK, about environmental, political, social and ethical issues leading to a regulatory regime for recombinant DNA (Krimsky 1982, Bennett et al 1986). Similar fears and questions are now voiced in the related but different context of genomics.

Genomics and the Human Genome Project

By the early 1980s researchers were beginning to search for and identify genes associated with particular diseases, such as the gene for cystic fibrosis and the gene for breast cancer. Initially such tasks were slow and tedious but the rewards were potentially great, and new technical advances began gradually to speed up the process. The ground was being prepared for achieving Haldane's predicted goal of an 'anatomy and physiology of the human nucleus'. By the end of the 1980s there was sufficient confidence to propose a Human Genome Project (HGP) to map and sequence all the human DNA, an enormously ambitious task given the technical resource at that time. It was then estimated that three billion base pairs would have to be sequenced, which would include perhaps 100,000 genes. The project would take 15 years and be carried out by an international consortium of laboratories at an estimated cost of $3 billion. As we shall see it was completed by 2003, two years ahead of schedule, and the number of genes proved to be far less than anticipated. The advent of the HGP placed genomics on the public agenda; genomics, amongst other things, deals with genes and their expression and so plays an important and increasingly important role in advancing the understanding of the causes of human diseases. President Clinton captured this vision in his 1998 State of the Union Speech when he said:

> In the 1980s, scientists identified the gene causing cystic fibrosis; it took nine years. Last year, scientists located the gene that causes Parkinson's disease – in only nine days! Within a decade gene chips will offer a road for the prevention of illnesses throughout a lifetime ... A child born in 1998 may live to see the 22nd Century.

Genomics will, it is predicted, be able to produce equally wonderful advances in agriculture, nutrition, criminology and other fields. That is the sunny side, is there a dark side to this vision?

The New Genetics: a Force for Good or Evil?

Haldane, it will be recalled, likened his vision for the future of genetics – now in part realised by the HGP – to nuclear power, 'a force for good or evil'. This was an apposite analogy. During the first four decades of the twentieth century physics was revolutionised, and by the end of the 1930s it was clear to the leaders that atomic power was a probability. The Manhattan Project turned this into a technical reality, and as Robert Oppenheimer put it, the scientists 'learned sin' (Snow 1981: 120). The moral distaste felt by some of the physicists who had worked on the Manhattan

Project, such as Maurice Wilkins, pushed them towards the life sciences and so strengthened the development of biophysics. Later, as we shall see, laboratories that were linear descendants of the Manhattan Project were instrumental in laying much of the groundwork for the HGP. A path, tortuous certainly, led from the Manhattan Project to the HGP. The achievement of the goals of the HGP when linked to the technical resources of biotechnology, and information science provide new sources of knowledge for awesome powers. Nuclear power made it possible to destroy civilisation, the new genetics makes it possible to change our genetic nature.

Biology and classical genetics in an earlier era provided us with concepts such as race, and practices such as eugenics that have led to dreadful social consequences. Eugenics has been a leitmotiv in the history of genetics, even Haldane still looked towards a 'scientific eugenics'. Not surprisingly, revisiting these issues in the light of the new genetics is uncomfortable. The Human Genome Diversity Project, proposed as a spin-off from the HGP, was shot down because it threatened political and ethnic interests. The eugenics of our era, might as Hilary Rose (1994) suggests, prove to be a 'friendly' free-market eugenics, as opposed to previous state-sponsored eugenics. Molecular biologist Lee Silver (1998) postulates that the market forces for genetic change in humans will prove so strong that in the future we might have two new social classes, the rich and genetically enhanced, and the poor non-enhanced. Stock (2002) also argues that governments will prove unable to prevent us 'choosing our children's genes', whilst the science fiction film *Gattaca* (Columbia 1997) presented a dystopian vision of a society in which that was the norm. Fukuyama (2002) who famously pronounced the end of history, revised his thesis a decade later, in the light of advances in biotechnology and the new genetics, to proclaim we were now facing a 'posthuman future'. On the broader economic front, critics such as Jeremy Rifkin (1998) are not convinced that growth in 'genetic commerce' will be an unmitigated blessing, and we witness in Europe and some Third World countries popular and governmental opposition to the introduction of genetically manipulated crops developed by US firms. The question of whether or not the geno-technologies are truly qualitatively different in their potential impact from other new technologies, such as for example information technologies, is frequently raised. Our position is that because they offer the possibility of changing Nature at its very heart, the germline, their assessment and regulation needs to be taken extremely seriously. It is not a question of exaggeration and overblown metaphors such as 'Frankenstein crops' or 'playing God' but a justifiable concern about crossing certain boundaries without the most thorough and publicly transparent examination of the consequences.

Structure and Distinctiveness of this Book

This book is organised around three related themes, following this brief introduction. In the first, accomplishing genomic research, which covers Chapters 2 to 4, we discuss the technical, cultural and philosophical issues surrounding the search for the 'book of life'. We look in detail at the worldwide attempt to map the entire make-up of the human genome, paying particular attention to the experience in the US and UK. In the second, commercialising the natural world, in Chapters 5 to 7, we focus on the genetically modified crops and food, and pharmacogenomics, and the rise of new genomics firms. This raises science and technology policy issues deriving from the Human Genome Project and their implications for patenting 'life' itself. The final part, mainly in Chapter 8, shifts the focus onto democratising involvement in decision-making in this key area, which affects everyone at a personal level. In particular we investigate the potential risks of the new genetics, and evaluate recent attempts, such as citizens' juries, to involve the public in discussing their implications. The book concludes with a discussion of the process of stabilising the HGP through black-boxing in order to develop a new genomics paradigm as the basis of a vision for the future. The discussion oscillates throughout between the local and the global reflecting the complexities of the issues involved, and in particular discusses developments in the UK, EU and USA.

This book has a number of distinctive features. It concentrates on one of the most controversial of the new technologies, which will impact on us all. Resources used in our study include the theoretical insights developed by the social studies of science and technology and science policy analysis (see *inter alia* Jasanoff et al, 1995), combined with results from a range of empirical work undertaken by the authors and colleagues using a variety of methodologies. The study of the many different aspects of the new genetics and society has benefited from recent advances in qualitative methodology as the issues raised to a large extent rest, as noted above, on the boundaries between selfhood and Nature. Hence most of the empirical research that underpins this work uses a variety of qualitative methods; focus groups, interviews, and participant and non-participant observation. Where quantitative methods such as large-scale surveys have been carried out, these have been helpful mainly in making the findings of the qualitative data more generalisable. In this way the book seeks to contribute to developing both the breadth and depth of existing research.

In Chapter 2, we show how much of the drive to establish the Human Genome Project was exterior to the field of human genetics. Certain groups, including scientists and bureaucrats, were able to mount a political lobby successfully, and enlist supporters in government circles, research agencies and media, in the teeth of opposition from many members of the biological sciences community. The scientific case for the megascience project was buttressed by organisational and national prestige, political, economic and health reasons. We also seek to demonstrate the importance of instrumentalities and technology in the HGP.

In Chapter 3 we examine the evolution and development of the HGP as it progressed from launch to completion in 13 years, two years ahead of schedule. The period from 1991–1997 saw steady progress in the establishment of mapping and sequencing centres dominated by public consortia that emphasised accuracy and the rapid public dissemination of data. The period 1998–2003 saw an intensification and speed-up of sequencing consequent upon the arrival of a privately funded rival. The ensuing struggle and competition throws up for study several important and economic issues that highlight stresses in the traditional ethos of science. In the following chapter we discuss the difficulties associated with managing the data explosion generated by the mapping project and its applications, including the issues of access, privacy and discrimination, especially in the context of the rapidly expanding number of Genetic Data Banks in the UK and abroad. We also focus in part on the attempts to deal with the data explosion by scientists working collaboratively using electronic forms of communication.

In Chapter 5 we explore how new genetic technologies cannot be divorced from the socially and politically co-constructed nature of social life by focusing on a related aspect of their commercialisation, the growing, international controversy over genetically modified food and crops, the role of multinationals such as Monsanto, and the Pusztai affair in the UK. We also discuss the ways in which the wider public can become engaged in such controversies. In the next chapter we discuss the process of globalisation and the role of the nation state, and their implications for the divide between the rich globalising North and poor South. We show how this affects not only their development, but also the distribution of health care, and initiates a process of commodification of the natural world.

In Chapter 7 we analyse the commodification and commercialisation of genomics. The emerging new models of innovation and technology transfer are described alongside the issues that they raise for public research institutions, in particular we examine the problems surrounding DNA patenting. The final section of the chapter looks at specialised genomics firms and the successes and failures of the business models that they have adopted. Chapter 8 shows the dangers and misunderstandings which arise from characterising the public as ignorant about the complexities of the potential risks raised by the new genetics, and how these might affect people in the future. We show that socially robust knowledge needs to be developed, and evaluate some of the new ways in which this can be facilitated. One recent method, the citizens' jury to involve the people of Wales in a debate about the introduction of genetic testing for common disorders into the National Health Service, is taken as a case study.

The book concludes by developing the arguments and insights discussed in earlier chapters that recognise that the new genetics is at once a global, multinational, multi-billion-dollar enterprise, and a very private and personal focus for concern. In the context of the completion of the HGP, the emergence of a post-genomic research paradigm, and the translation of biology into 'big science', we explore the ways in which these complex and multifaceted social processes are likely to unfold in a variety of policy arenas. In doing so we focus on the visions for the future that is the basis of wide-ranging debates in Britain and the USA.

Chapter 2

The Hunt for the Holy Grail: Compiling the Book of Life

'The total human sequence is the grail of human genetics.' (Walter Gilbert cited by Cook-Deegan 1994: 88.)

In this chapter we examine the origins of the Human Genome Project (HGP), and how political, scientific and technological forces shaped it. In particular we seek to show how this megascience project, seeking to sequence the entire human genome, was dependent on, and co-evolved with, developments in the technologies of practical research, which we term instrumentalities (Price 1984).

An HGP was not on the research agenda or wish list of most biologists in the early 1980s, certainly it was not a necessary consequence of the then current state of human genetics, the scientific field which one might imagine to be the one with the most to gain from it. Yet by the time of the announcement of the rough draft of the human genome in 2000 it was being hailed as one of the greatest achievements of all time alongside '...Bach's music, Shakespeare's sonnets and the Apollo Space Programme...' (Richard Dawkins cited in Davis 2001: 241). How then are we to explain the emergence of such a unique biological mega-project?

To appreciate how the Human Genome becomes a possibility one needs to study several different strands of activity, none of which in themselves are a sufficient explanation, but together form a causal nexus which make its emergence seem a logical and predictable development of the New Genetics. The most important of these causal strands are:

- The Politics of Science, particularly US science
- The role of instrumentalities in science
- The evolution of human genetics

We will explore and analyse these in this chapter.

Metaphors and the Human Genome Project

To launch a research endeavour on the scale of the HGP is no easy task for, in addition to the obvious scientific issues, it contains enormous political and economic dimensions. Thus, the scientific proponents of the HGP had to enrol into their network of supporters people who were not themselves scientists. Such people and groups had to be persuaded that HGP was something meaningful to them. The use of metaphor played an important part in their enrolment, we find the HGP being variously described by its early proponents as 'the holy grail of genetics', 'the moon shot of genetics' and the 'book of life'.

It is worth briefly comparing and contrasting these metaphors, since they alert us to the scientific and technological aspects of the programme, and how they interrelate.

The holy grail was, according to some legends, the chalice or cup used by Jesus at the Last Supper. Symbolically, the quest for the holy grail represented the search for the secret of life, the supreme secret (Ross and Ross 1926). Epistemologically, the metaphor of the HGP as a holy grail is flawed, in the eyes of some critics, on the grounds of reductionism and biological determinism; furthermore such critics maintain the scientific search is, in any case, a sideline of a more fundamental crusade for power and profits (Lewontin 1991).

It can be argued that the moon shot metaphor is a no more accurate description. The moon shot was a technological programme undertaken for political prestige, rather than for scientific reasons (which is not to say there has not been some scientific fall-out from it, but that was not its raison d'etre). Once the Apollo programme was completed the next obvious technical steps such as a moon base or Mars trip, which provided no immediate political advantage, were cancelled. The HGP is not actually comparable to the moon shot as a political prestige project, although as we later discuss, it was hijacked by presidents and prime ministers at the time of the completion of the rough draft. Certainly, however, we can say that the technological push behind the HGP outweighed the scientific pull. So, if the HGP did not have an overwhelming scientific rationale, and lacked the obvious prestige and excitement of the moon shot, why was it eventually adopted?

The view presented here is that it can be understood in large measure in the context of the developing relationship between molecular biology and the biotechnology industry, underway since the 1970s. Almost without exception the leading actors in America were molecular biologists with strong links to the biotechnology industry, or academic or research administrators looking for a 'big science' new generation biotechnology bandwagon on which to hitch their institutional ambitions. These people were ultimately able to build a sufficiently strong lobby effort to win the promise of large public funds from funding agencies such as the National Institutes of Health (NIH); the process involved massive hype about potential medical benefits. Debate over whether or not the HGP was good science, routine or applied science, or technology turns out to be rather sterile, because from the perspective presented here, the HGP ought to be primarily seen as an instrumentality development programme.

For that reason our preferred metaphor for the HGP is the 'book of life' (Horizon 1988). Books, as we know, come in many kinds. At one extreme we find stories, and at the other we have handbooks and catalogues. The HGP book of life is not the story of life, an extreme reductionist view, although it will help science write that book. Rather it is to be seen as more a handbook or catalogue of our gene sequences and their chromosomal positions; Bodmer (1993) refers to it as the 'telephone book of life'. Some telephone book; Rosenberg (1998) calculated that it would be equivalent to a double stack of telephone directories a mile and a half high! Of course, it is unlikely to ever be a real printed book, for it is stored in vast computer databases and gene banks able to provide genetic information for scientific research, technological development and medical treatment. Shortly after the draft sequence was published a copy of the human genome sequence was placed on a CD-ROM, and given away by some magazines as a souvenir! That was, of course, part of the celebratory hype at the time. Thus the end product of the HGP is not a holy grail – the secret of life, nor a technological spectacle like the moon shot, it is, as Sydney Brenner put it 'a tool' for scientific research (Davies: 248); the metaphor of a tool is explored in more depth in our later discussion of instrumentalities in genome research.

Political Origins

Most of the early popular accounts of the origins of the HGP overemphasise the importance of the scientific precursors, especially human genetics. In fact a better case can be made for regarding the HGP as a technological vision and result of political lobbying by certain research bureaucracies. Cooke-Deegan's excellent and exhaustive account of the politics of the origins of the HGP powerfully supports such a view (Cook-Deegan 1994). He argues that while it would have been logical for the HGP to have emerged from human or medical genetics, that is not, in fact, how it happened. None of the original protagonists, he claims, was a specialist in human genetics. He argues that the HGP was initially based on a technological vision of systematically applying the tools of molecular genetics to sequence the entire human genome, although as the idea of the project became established, human geneticists were able to redefine the goals. Cook-Deegan also demonstrates the importance of building appropriate bureaucratic structures for mega-projects and the role played by politics and associated intrigues.

In various places in this chapter, and in the next, we use timelines from which we are able to obtain some of the historic highlights discussed in our analysis, in the knowledge that more detailed histories exist elsewhere, for example Cook-Deegan (1994), Davis (1990), Davies (2001). Table 2.1 provides a timeline of some of the key elements in the origin of the HGP some of which we will expand upon to explore forces leading to its construction.

One of the first persons to propose a programme to map and sequence the human genome was the American molecular biologist Robert L Sinsheimer, chancellor of the University of California, Santa Cruz (UCSC) in the late 1970s and 1980s. University chancellors are expected to think up fund-raising schemes to increase their institution's prestige. One method of doing this is to develop big science projects. Having recently lost out on a large-scale astronomy project Sinsheimer wondered whether perhaps a big science project was needed for biology – one which could be housed in Santa Cruz, funded by an appropriate foundation.

Table 2.1 Origins of the Human Genome Project (based on Cook-Deegan 1994, and *Science* 16 February 2001)

1985	Robert Sinsheimer hosts meeting to discuss feasibility of sequencing the human genome at University of California Santa Cruz Sydney Brenner urges European Union to undertake a human genome sequencing programme
1986	US Department of Energy (DOE) Santa Fe Meeting to discuss plans for sequencing human genome Renato Dulbecco writes Science editorial promoting sequencing the human genome
1987	Walter Gilbert plans a private scheme to sequence human genome, Genome Corp. DOE presents plan to lead a US sequencing effort
1988	NRC Report endorses idea of HGP NIH decides to have a sequencing programme, challenging DOE for leadership Office of Technology Assessment Report on HGP US House Committee on Energy and Commerce Hearings on NIH-DOE collaboration on HGP Human Genome Organisation (HUGO) founded NIH and DOE sign memorandum of understanding to collaborate on the HGP James Watson appointed to lead NIH genome research programme NIH establishes National Centre for Human Genome Research
1989	UK genome programme launched by MRC International dimension emerges as programmes launched by European Commission, Japan and others Joint DOE-NIH committee on ethical, legal, and social implications of the HGP
1990	NIH and DOE publish first joint 5-year plan 1 October declared official start of the HGP by NIH and DOE

'The characteristics of Big Science projects … was that they provided a facility that would be essential to further advance in the field. Biology did not seem to need a comparable facility. What biology needed, however, was a massive information base – a detailed knowledge of the genetic structure of several key organisms, including for obvious reasons – man' (Sinsheimer, 1994: 264–265). Sinsheimer, from his scientific background, knew that genetic mapping and sequencing were developing in the direction that made genome sequencing plausible, although nobody at that moment seemed to be making a case, as he was about to, for sequencing the entire human genome. He optimistically estimated that to do this would need 'a building, equipment, and endowments adding up to perhaps twenty five million dollars' (Sinsheimer 1994: 266). The big science project for UCSC could, thought Sinsheimer, be an institute devoted to human DNA sequencing. In a letter to David Gardner the President of the University of California he wrote such an institute could be '… a noble and inspiring enterprise. In some respects, like the journey to the moon, it is simply a "tour de force"' (Cook-Deegan 1994: 82).

He felt that UCSC would stand little chance of Federal funding in competition against the heavy hitters like Stanford, MIT etc., and so decided on seeking funding from a foundation. To explore the feasibility of his concept he convened a small workshop whose members were '… representatives of the most active sequencing groups, researchers interested in the development of associated automated instrumentation and computer techniques' (Sinsheimer 1994: 267). The workshop held on 24–26 May 1985 is regarded a landmark in the history of the HGP. John Sulston, the English sequencing expert later recalled 'I felt amazed that we were all sitting there discussing making an attack on the human genome' (Sulston and Ferry 2002: 59). Amongst the other attendees were Walter Gilbert, who received a Nobel Prize for his DNA sequencing technique and Leroy Hood, who pioneered the development of the DNA sequencer.

The conclusions of the workshop were briefly that:

> (1) A genetic map of the human chromosomes providing well-defined markers (polymorphisms) at reasonable spacings along all of the chromosomes, to use as reference points, could be developed (in collaboration with outside groups) by a staff of perhaps twenty people in a two-to-four year period.
>
> (2) A physical map of the human chromosomes providing a linearly ordered set of cosmid-size (thirty to forty thousand bases) DNA fragments could similarly be developed by a group of twenty people in two-to-four years.
>
> (3) A complete nucleotide sequence map of the human chromosomes is not presently feasible with reasonable effort. Sequencing a few percent of the genome around selected markers and in carefully chosen regions is feasible, with a group of thirty people working over ten years. The availability of the sequences would undoubtedly be of great value. At the present time, it is quite reasonable to anticipate advances in and automation of sequencing technology such that the sequencing of the next few percent could be done with one fifth or one tenth of the man-years effort.
>
> (4) There was general agreement that a centralised effort correlating genetic, physical, and sequence mapping, promoting the development of improved technologies, and actively fostering the application of this knowledge and approach to specific problems in human genetics, development, and physiology would be of great value.
>
> (Sinsheimer 1994: 268)

At this point we need to point out the differences between 'genetic, physical and sequence mapping' listed in item (4), without getting involved in detailed technicalities. Genetic maps are a relatively old tool dating back to the early days of classical genetics. As long ago as 1913 A.H. Sturtevant, working in T.H. Morgan's lab, produced the first map, showing the linear sequence of six sex linked genes on the X chromosome of *Drosophila* (Dunn 1991). Over the years since then geneticists refined the linkage methodology which allowed them to place genes or DNA markers in order along chromosomes. Usually when it is announced that the gene for a particular condition or trait has been discovered it means that researchers have been able to place it on a genetic map. We then know its position relative to other genes or DNA markers, however that in itself cannot tell us the gene's structure, though obviously knowing approximately where a gene is along a chromosome is an important step in that direction.

Physical maps were a later development, the type developed in the HGP are called contig maps. These are made from overlapping cloned fragments of DNA arranged in the order in which they occur along a chromosome. The DNA fragments can be cloned in bacteria, termed BACs, or yeast, called YACs. The physical and genetic maps can be compared to locate genes on particular fragments, this provides the basis for a process to create a genomic map. Then as Sulston vividly describes '...you just don't know the gene as an abstract point on a diagram; you know that a particular colony of bacteria in your freezer contains that gene within the clone it harbours, and so the sequence of the gene is within your reach. (Sulston and Ferry 2002: 41–2). Meaningful sequencing of the whole genome only becomes possible after the creation of high quality physical maps, '...the sequence is in one sense the ultimate map, it is also much more than that: it is also the biological information itself' (Sulston and Ferry 2002: viii).

Sinsheimer was never able to raise the necessary funding for his Institute. Later he wrote, 'Somewhat naively, I believed that the project...would surely seize the imagination of anyone with even a rudimentary scientific bent. Curiously, it did not. I also believe that the proposal would have been given greater credence and a better hearing had it been put forward by a more prominent, established institution – a Caltech, a Stanford, a Harvard. UCSC was an undistinguished spot on the map of biological research. Ideas should be evaluated purely on their merit, but in the real world, in the battle for attention and credence, that seldom happens' (Sinsheimer 1994: 268).

However, Sinsheimer did circulate the workshop report, which in Sulston's words 'added to a groundswell that was beginning to emerge in favour of making a concerted approach to the human genome.' (Sulston and Ferry 2002: 60)

Whilst Sinsheimer failed to get any foundation to support him, Walter Gilbert looked towards the private sector. Joel Davis (1990: 147) says 'Walter Gilbert is good at many things. Three of them are doing science, creating companies, and getting people riled up.' Gilbert might be considered a type species of the late twentieth century new biologist able to bestride the fields of molecular biology blue-sky research and commerce. He shared the 1980 Nobel Prize for chemistry for developing a technique for sequencing DNA, and in 1982 became CEO of Biogen, a pioneering biotechnology start-up company – leaving an professorial post at Harvard to do so. By late 1984 he had moved back to Harvard to chair its biology department (Cook-Deegan 1994: 87). Gilbert, as we have noted above, attended the Santa Cruz Workshop, becoming over the next year an enthusiastic missionary for the HGP. Through his wide network of scientific colleagues he was able, as Cook-Deegan says, '...[to] carry the ideas from Santa Cruz into the mainstream of molecular biology' (Cook-Deegan 1994: 88); and in the process enthusing some leading individuals, including James Watson. He also appeared in the press and on TV as an advocate for the HGP, and in so doing helped to take the idea of the HGP into the public realm.

At the 1986 Cold Spring Harbour Symposium, Gilbert had estimated that sequencing the human genome's bases would cost $3 billion, at one dollar per base. Not surprisingly a project on such a scale was seen by many biologists as an expensive threat to other publicly funded biological research, a point which we explore later.

Gilbert further irritated much of the scientific community by suggesting in 1987 that he might privatise the HGP through a start-up company called Genome Corp. This would '...construct a physical map, do systematic sequencing, and establish a database. The business objectives included selling clones from the map, serving as a sequencing service, and charging user fees for access to the database. The market would be academic laboratories and industrial firms, such as pharmaceutical companies, that would purchase materials and services from Genome Corp' (Cook-Deegan 1994: 89). In a sense this idea merely developed an existing trend of biological services industries that enabled researchers to avoid, at a price, routine tasks and get on with their creative research. Unfortunately, from his point of view the timing was wrong. Genome Corp. failed for two main reasons to raise the necessary capital. First, after the 1987 stock market crash, biotechnology investment was at a low ebb, investors having become sceptical of biotechnology's claims. Second, because of impending competition from public research, whose data would be freely available. A decade later, as we will see in Chapter 3, the entry of Celera into human genome sequencing would raise the controversy over public versus private research in the HGP to a new level of intensity.

At the time Gilbert proposed his Genome Corp it was by no means certain that public funding on the necessary scale, tens to hundreds of millions of dollars annually, would be forthcoming for the HGP. The obvious funding candidate in the United States might seem to be the National Institutes of Health (NIH), responsible for the world's largest public bio-medical research funding. However, it initially failed to see the proposed project as a worthwhile endeavour.

It was Charles DeLisi, Director of the Office of Health and Environment Research (OHER) in the US Department of Energy, a descendent of the Atomic Energy Commission, who first put funding of the HGP on the US Federal agenda. In 1985 DeLisi had read a draft report of a meeting held in December 1984 at Alta, Utah sponsored by the DOE and the International Commission for Protection Against Environmental Mutagens and Carcinogens. The Alta meeting was about the need to find sensitive assays for human heritable mutations, not about a HGP. However, DeLisi maintains that it was through reading the report of the Alta Meeting that he first had the idea that the DOE might organise a human genome project (DeLisi 1998). The Alta meeting therefore proved to be a bridge between the HGP and the Manhattan Project (Cook-Deegan 1989), which is discussed below.

In March 1988 DeLisi had organised a meeting in Santa Fe to explore the idea of a HGP. It was there that Gilbert famously declared 'The total human sequence is the grail of human genetics' (Bishop and M. Waldholz 1990: 218). During the week of the Santa Fe meeting Renato Dulbecco, a Nobel Prize winner published an editorial in *Science* (Dulbecco 1986), in which he argued that sequencing the human genome would speed up cancer research. Victor McKusick has been quoted as saying that '...perhaps more than any other single factor, the editorial galvanised the scientific community and even the public, and also polarised the scientific community to some extent' (Bishop and Waldholz: 218). Cook-Deegan succinctly summarised the situation surrounding the origins of the HGP: 'Sinsheimer convened the first meeting dedicated to discussing whether or not to sequence the human genome, and DeLisi laid the first stones in its bureaucratic foundation, but Dulbecco was the first to publish the idea in a large-circulation journal aimed at the entire scientific community' (Cook-Deegan 1994: 109).

When James Wyngaarden, then Director of NIH, heard in mid-1986 about the proposed DOE genome programme he was shocked, remarking it seemed '...like the National Bureau of Standards proposing to build the B-2 bomber' (Cook-Deegan 1994: 89). This was an understandable comment, but not completely fair. DOE in its various earlier institutional incarnations had been carrying out genetic research since the days of the Manhattan Project and the A-bombing of Hiroshima and Nagasaki. Within the DOE a genome project could be regarded as a spin-off from its earlier military interest in the effect of radiation on the Japanese survivors and their descendants. There had in fact been a long-term programme of research since the mid-1940s, leading to great technical expertise on mapping, and by the 1980s, this involved molecular genetics. A move to DNA sequencing was seen as a logical progression, especially as some of the DOE's other research fields were beginning to look dated.

As the DOE built up a case for its genome programme (presented in great detail by Cook-Deegan 1994: Chapter 7) several illuminating arguments emerged.

Its in-house expertise in human radiation genetics and technical knowledge, which included computing, robotic laboratory instrumentation and bioinformatics databases, would be germane to the HGP. Participation would provide the opportunity to use genomics to raise the reputation of DOE biological research to the world class enjoyed by its high energy physics research, at a time when biology seemed to be moving towards a pre-eminent position in science. Further, it was perceived that a HGP would allow the DOE to present a less militaristic image; one internal memo wrote 'Oppenheimer's statement "I am become death, the destroyer, the Destroyer of Worlds", gives way to 'the National Laboratories are become the ultimate advocates for the study of human life'. A project in which DOE would become 'a DNA-centred mechanism for international cooperation and reduction in tension' (Cook-Deegan 1994: 97). Finally, in the light of the running down of the Cold War, the HGP could provide some kind of financial and institutional safeguard for the Los Alamos and Lawrence Livermore national laboratories, whose work was mostly military. This was put rather bluntly in May 1987 by the senator for New Mexico, Peter Domenici, concerned about the future of benefits brought to his state by the DOE National Laboratories, 'What happens if peace breaks out?' (Cook-Deegan 1994:104).

Given the role that the NIH was to play eventually in the HGP it now appears a little strange to say the least that the initial Federal push for the HGP came from the DOE and not the NIH. The reason why the DOE moved towards the HGP more rapidly than the NIH was because its leaders felt that the DOE needed it. On the whole the biologists associated with the NIH were happy with the way things were. Research on gene mapping was funded by NIH as small science, in which individual research groups would be funded on the basis of their proposals, i.e. it was a 'bottom up system'. What the proponents of the HGP had in mind smacked to many scientists of a top-down approach, threatening their scientific autonomy and creativity. The plans of the DOE however, disturbed some biologists since it posed a boundary problem. It didn't seem right to some molecular biologist supporters of the project to leave the genome project to the DOE, which they saw as an organisation of physicists trying to do biology.

Before the NIH could create a genome programme it needed to secure funding and the support of its biological community, and at the time this was by no means straightforward. According to Cook-Deegan, James Wyngaarden deserves credit for ensuring funding would be forthcoming.

> Wyngaarden knew that most NIH policy was determined by the appropriations process, and regarding the genome project, he focussed on this objective' (Cook-Deegan 1994: 143). Having secured the budget he created an appropriate bureaucratic structure for the genome project within NIH by establishing '... the Office of Human Genome Research to co-ordinate efforts during 1988 and 1989 and convinced his departmental superiors to create the National Centre for Genome Research. This gave the human genome project a more permanent home in the NIH bureaucracy, and 'centre' status assured independence by conferring the power to directly to disburse funds. Without special attention from him, it is unlikely NIH would have moved with as much dispatch. Wyngaarden did indeed react to the DOE initiative, rather than generate the idea for program, but his task was no smaller.
>
> (Cook-Deegan 1994: 147)

As we have already indicated, many biologists didn't welcome the HGP. The debate basically swung around the issue of the HGP as a big science. Here was a project which might cost up to $3 billion over its lifetime, and to biologists that seemed an awful lot of money. Although compared to other contemporaneous big science proposals, such as the Superconducting Supercollider, at $8 billion, and the International Space Station, at $40 billion, it could be regarded a small big science. Walter Bodmer thought so and enthused that the HGP was '...the most exciting science... more so than all these space shots and Hubble telescopes, that have slight distortions in them so you can't see clearly... that distortion would have paid for the HGP ... the Supercollider could pay for a few HGPs' (Lee 1991: 238).

Thomas F. Lee (1991: 238) wryly observed that some proponents, like Renato Dulbecco simply denied the analogy of big science as being inappropriate. Leroy Hood argued 'Rather than big science, it is interdisciplinary science.' (Lee: 238) Others however, revelled in it. Charles Cantor, of the DOE, argued that unlike other big science projects '...this project is guaranteed to succeed...it is not basic research in any sense...it is engineering to provide a tool that will be used by people in basic research' (Lee 1991: 238).

James Watson thought there was no point in criticising the project for being big science because

> well the human genome is big...I never thought that big was necessarily bad. We want it to be big enough so that the cost can be brought down to a reasonable level...[and] we want to get it done in a reasonable period of time...
>
> (Lee: 238).

In recent years the term big science has been replaced in some policy circles, especially in OECD, by megascience. Its policy analysts distinguish two kinds of megascience projects. Those with a central facility such as CERN or the Space Telescope, and those with a distributed facility such as the HGP. In the latter the researchers can be spread over a numbers of geographical sites. It has been reported that 'Data are the glue that holds distributed megaprojects together…effective data collection, management, evaluation and distribution are crucial to [their] success' (Ratchford and Columbo 1996).

Not everyone in the biological sciences community was as sanguine as Watson. There were several major concerns within the scientific community. First funding the project would mean less for other biological research, and second, that it was a technological rather than scientific project – i.e. it was not good science. Lee summarises the arguments as to whether or not the HGP would result in good or bad science (Lee 1991: 242–245). The main argument against the HGP allowing good science is that poor science would result because investigator initiated research, i.e. bottom-up research, has proved itself to be the most successful research approach. Whereas, a top down approach would be bureaucratic, targeted research in which the scientist is told what to do. Such an approach tends, it was argued, to attract mediocre researchers, technicians rather than creative scientists. Supporters of the view that the HGP is good science argued that it is not true that bottom-up research is the only valid approach and that top-down targeted research is better under certain conditions, and mapping and sequencing is such a situation. Furthermore, given larger scale funding creative researchers could be freed from tiresome repetitive work by the automation of laboratory technique, and therefore good researchers would be attracted to the HGP. Although not everybody was convinced (Lowotin 1991), a series of discussion meetings, combined with the bureaucratic and funding success mentioned earlier did lead to some degree of consensus allowing the debate to move on to other issues, such who was to be the lead US agency – NIH or DOE?

By 1987 the White House's Biotechnology Science Co-ordinating Committee (BSCC), the National Science Foundation (NSF), the Congressional Office of Technology Assessment (OTA), and the DOE all had committees studying the matter of who should lead the Human Genome Project. James Watson in mid 1987 made a key intervention raising two important points:

> (1) Bureaucrats must not be in charge of co-ordinating Genome Project work done by the agencies involved. The co-ordination must be done by scientists. (2) It would not be possible to carry out such a major scientific and technical project without a lead agency. There's only one genome, and we need one lead agency.
>
> (Davis 1990: 138)

The appointment in 1988 of Watson, by Wyngaarden, as head of the new NIH Office of Human Genome Research was seen as a smart move helping to place NIH in the ascendancy.

> DOE and NIH would eventually get together and thrash out a Memorandum of Understanding about co-ordination of their genome mapping programs. But when Watson

> agreed to accept the position … the question of which was the de facto lead agency for the Genome Project was effectively settled … it is the National Institutes of Health.
> (Davis 1990: 140)

Congress set up for the fiscal year 1988 through 1990 separate appropriation bills for genome work in DOE and NIH. 'The size of the budgets ($58.5 million at NIH and $26 million at DOE, was an implicit measure of relative power' (Cook-Deegan 1994: 147).

Interest in the HGP was not, of course, confined to the United States. By 1989 an international dimension was emerging with European countries, such as the UK and France, and Japan launching their own genome programmes.

In early 1990 the NIH and DOE published a five-year plan for the HGP, whose goals were very similar to those produced by Sinsheimer's workshop in 1985. They included:

- Completing a fully connected human genetic map
- Creating a physical map, assembling STS maps of all human chromosomes
- Improving sequencing methods, seeking to reduce cost to 50c per base pair
- Start mapping and sequencing model organisms, other species such as *Escherichia coli*, and *Caenorhabditis elegans* to gain necessary experience for human work
- Develop informatic techniques, data collection and analysis (US Human Genome Project 1990)

During this period it was recognised that it would be premature for technical and economic reasons to actually begin sequencing the human genome. The clock for US HGP was officially started on 1 October, 1990.

Instrumentalities

The history of science has often been presented primarily as a history of ideas, and of the development of theory by outstanding individuals. Such an approach is, of course, one-sided; while creating and promulgating myths of the isolated genius and profound yet abstruse knowledge, it also tears science out of its broad social context. As a result science is commonly seen as other-worldly, incomprehensible and generally separate from the normal activities of making a living. However, there are certain processes integral to the way science is done which tie it to the so-called 'real world' of industry, business and politics. Indeed, without these linkages science can't be done effectively, the progress of the HGP is an excellent example of this process. Furthermore, the more we can understand the nature of such connections the better chance we have of appropriating the results of scientific research through technological transfer and innovation.

The historian of science Derek de Solla Price, argued that interesting science/technology interactions occur in fields where '…basic and applied research are inseparably linked to technology by the crafts and techniques of the experimentalist and inventor' (Price, 1984: 10). When the role of experimentalists in scientific progress has been studied by science historians their emphasis has usually been on the scientific instruments devised and used for observations, measurements and experiments. While it is true that scientists' experiments are designed to test and confirm or falsify theories, in the process of so doing they find out an awful lot of information about matter, effects, and techniques that may well be applied and used in places other than the laboratory. As Price (1984: 12) shows; '...from time to time these neat effects would be raided by technology whenever market potential became visible. It is that that rather strongly links the techniques to both scientific theory on the one side and industrial application on the other.'

So strongly did Price believe in the importance of this neglected aspect of scientific progress that he argued;

> We need a new term for these important techniques that help make new science. It will not do to call them instruments. Although the telescope fits this category, our term must let us include parts of the experimental repertoire that are labeled 'effects', such as the production of voltaic electricity, or the photo-electric effect, and such things as Cerenkov radiation or nuclear magnitude [sic] resonance. We must also include chemical processes, such as polymerisation and Lowry's method for protein determination, and biological processes, such as recombinant DNA that lead to genetic engineering. I advocate the use of the term instrumentality to carry to general connotation of a laboratory method for doing something to nature or data in hand.
>
> (Price 1984: 13)

Scientific research methods have expanded greatly since Price wrote this and it now seems reasonable to expand his list to include reagents, analytical methods, software and so forth.

It is because science cannot be done without these instrumentalities that attempts to understand their role in the process and progress of science should be encouraged. However, to concentrate only on this aspect would be to perceive only part of their social function. The role that scientific instrumentalities play in transforming industry, and how they come to do it, has been ignored even more than their role in advancing science (Rothman, 1994).

It is possible that the importance of the role played by instrumentalities varies between scientific fields; Price (1984: 14) thought that instrumentalities appeared to be particularly dominant in what he termed the 'new biology'. In the general field of biotechnology, there are a variety of techniques, including instruments, that clearly play a major part in the development of both of the technology and of the science. Certainly the role of instrumentalities in the HGP is paramount, indeed, the human genome sequence itself can, as we shall see, be considered an instrumentality. The instrumentalities of the HGP need not be studied simply as things in themselves but can be studied as co-productions within evolving socio-technical network complexes of the HGP. Hilgartner (1994: 305) using the language of actor network theory saw the task of the HGP as a

> heterogeneous engineering problem of immense proportions...[entailing] building a network – of researchers, techniques, organisations, laboratories, databases, biological materials, funding sources, political support, and so on – that can operate at high 'throughput' and produce maps and sequence data with few errors.

The Human Genome Project as Technology

The original supporters of the HGP seem generally to have been those who understood the technology and its potential, commercial as well as scientific. Initially, it was thought that effective use of this technology called for a big science approach. As we have seen arguments were made showing how the HGP had parallels in the high energy accelerators of particle physics, programmes and facilities that could only be made by national efforts and international collaboration. Although a large amount of money was being proposed for the HGP, by the standards of NASA or the Strategic Defence Initiative, it was relatively limited. A big science ethos was not one which matched normal research practice in biology; it was thought during the initial discussions of a strategy for the HGP that large centralised facilities would be the preferred laboratory form. As it turned out, the nature of the mode of research did not necessarily lead to the kind of big science set-up associated with big machines and instruments of particle physics and astronomy. It proved possible to hack out space for diffused networks of researchers, for example, the cDNA approached funded by the UK Medical Research Council are associated with a centralised resource unit, HGMP Resource Centre, which provides a series of technical services for small science researchers (Glasner et al 1995). Nevertheless, arguments for the industrialisation of research proved persuasive; sequencing was repetitive process which could be automated and carried out by technicians in a production line atmosphere. As the HGP developed, these began to be established, one of the first being Genethon in France (Abbott 1992, Cohen 1993). We will later describe others involved in the dash to complete the draft sequence.

There is little doubt that the original version of the HGP was unpopular with many biologists who thought it would be a diversion from more important scientific research. On the other hand, Sydney Brenner, whose Cambridge laboratory reputably had more experience of sequencing than anyone else at the time of the inception of the HGP, '...couldn't understand why people are not reaching out gratefully for this new technology but are being carried screaming into it'. Brenner also remarked that

> In biology ...[in] a strong sense all answers exist in Nature, all we need is the means to look them up, and that's what techniques give us ... In general, techniques have been absolutely important. We couldn't have got anywhere without them ... I'm very keen on technique but my colleagues aren't.
>
> (Horizon 1998)

In a similar vein, David Botstein, who pioneered the use of restriction length polymorphisms (RFLP) for genome mapping, complained that

> ...the genome project in particular suffers from that people are uncomfortable with research projects designed to improve technology as opposed to a research project designed to extract a few facts ... it's hard to convince people of the value of technology.
>
> (Horizon 1998)

Brenner and Botstein, who have both made a series of outstanding scientific discoveries, represent that faction of biological research wedded to the notion of the profound necessity of technique, that is instrumentalities, in the advancement of knowledge.

The Relationship between Theory and Experiment

The asymmetry of prestige between the theoretician and experimentalist in science is rather bizarre when one considers their mutual interdependence. Is it that experimentalists are less pure, proletarian rather than aristocratic, more likely to be tainted by the dirty world of production? There are, no doubt, many complex sociological and psychological factors at work, certainly it is often possible to distinguish individual scientists as either theoreticians or experimentalists, though this may be easier in physical than biological sciences. Interestingly enough, a number of Nobel Prizes in fields relevant to the HGP have been awarded to scientists for what we would argue are instrumentalities, for example, Sanger and Gilbert for work on sequencing, and Mullis for inventing the polymerised chain reaction (PCR) for DNA amplification.

When one studies the invention and development of instrumentalities, it is not uncommon to find a movement from the laboratory to industry and back to the laboratory again. Scientists such as Mullis invent laboratory techniques then, as the techniques diffuse through other laboratories and as a demand builds up, they are commercialised, in purer or more efficient forms, by specialised instrumentality manufacturers and sold back to laboratory scientists.

The commercialisation of laboratory technology, therefore, may well aid scientific progress, however, it can also create new tensions within the scientific community. The tempo of research is increased and the balance of power with respect to competitive edges changes. Scientists who can afford these commercial instrumentalities have an edge on those who cannot; that is the blunt message of many of the advertisements that embellish the pages of scientific journals. Even a cursory perusal of advertisements for laboratory equipment in scientific journals such as *Nature* shows the relationship between these products and research. Those working without the benefit of the latest instrumentalities in a given field are presumably finished as serious players in that area of research!

The contradiction between the opportunities offered to researchers by these developments and the limited funds to buy them has not gone unnoticed. Melvin Schindler (1992: 1423) observed:

> The contrast between the great opportunities for inquiry afforded by these instruments…and the limited funding provided by government to acquire these expensive tools is remarkable. There can be no doubt that technology is the engine for scientific advance. This is not to demean the importance of a good idea, but is the quality of the scientific tools that raise the level of questions and efficiency of the experimental approach.

GCUL

GENE RESEARCH
Improve Your PCR Process!
Reach the finishing line faster
Beckman Products optimise accuracy with automation and operational simplicity before and after the amplification step. This assurance of lower cost, error free performance is the Beckman Plus that will help you reach the finishing line faster.

From the outset the HGP was presented as technology dependent (DOE Office of Health and Environmental Research 1987); it was recognised that the technology available at the onset of the programme was insufficient to carry it out with the time and other constraints envisaged. Many of the early policy documents contained technology 'shopping lists'. Table 2.2 (OHER 1987) shows one such list. An important part of the HGP was the development and improvement of such items.

Table 2.2 Desired Technologies for Sequencing the Human Genome

(1) Production of DNA fragments containing 100–1000 kilobases
- Chromosome separation
- Restriction enzymes
- Separation and purification of large fragments
- Large insert cloning

(2) Automated DNA handling, mapping and sequencing
- DNA preparation
- DNA cloning
- Physical, restriction fragment and genetic mapping
- Chemical, physical and enzymic sequencing

(3) Data storage and analysis
- Immediate data entry for uniform notation
- Efficient search with cross referencing and access to other data banks
- Rapid data distribution
- Parallel or concurrent processing
- New algorithms for analysing and interpreting DNA and protein sequences

(4) Detection and analysis of DNA, RNA and protein at very low levels
- Single molecule analytical methods
- Methods of detecting large numbers of DNA fragments simultaneously

Similarly it was recognised by industry that commercial opportunities for selling more instrumentalities might emerge from the HGP (Fox 1991). The commercial development of sequencing technologies was, for example, regarded as particularly important; later we'll explore the interplay between public R&D and commercial developments in sequencer development. As the HGP progressed through the 1990s the details of technology may have changed but the perception of the project as technology dependent strengthened.

Table 2.3 provides a timeline of some of the more important HGP instrumentalities. From it we can see that by the time the HGP commenced there had been about two decades of necessary instrumentality development to hand; such as, restriction enzymes, DNA cloning techniques, sequencing techniques and DNA database developments. We can also see that during the 1990s that these were enhanced and improved, and significantly, great advances were made in computational methods and bioinformatics.

Table 2.3 Timeline of Important HGP Instrumentalities (*Science* 291, 1195, 2000)

1970	Restriction enzymes isolated: Hamilton Smith and Kent Wilcox isolate the first restriction enzyme – a protein that cuts DNA at specific sites defined by the base sequence. All of molecular biology uses restriction enzymes.
1973	DNA cloning developed: Stanley Cohen, Herb Boyer et al show that DNA joined to a plasmid molecule can be replicated in bacteria – the foundation of all DNA cloning work.
1977	DNA sequencing methods developed independently by Sanger and colleagues at Cambridge, and Walter Gilbert and Alan Maxam at Harvard.
1978	Sanger and colleagues develop use of thin acrylamide gels for improved DNA sequence analysis.
1980	Shotgun method developed by Sanger and colleagues developed to prepare templates for DNA sequencing.
1983	Cantor and Schwartz develop pulse field electrophoresis.
1985	Kary Mullis and colleagues at Cetus Corp. develop the polymerase chain reaction (PCR).
1986	First automated DNA sequencing machine: Leroy Hood, Lloyd Smith and colleagues at Caltech.
1987	Applied Biosystems Inc. market first DNA sequencer, based on Hood's technology.
1989	Olson et al propose new mapping strategy using sequence-tagged sites (STSs).
1990	Capillary electrophoesis developed separately by Lloyd Smith, University of Wisconsin; Barry Karger, Northeastern University; and Norman Dovichi, Alberta University.

1991	GRAIL pioneering gene-finding programme developed by E. Uberbacher, Oak Ridge National Laboratory.
1993	Mega-YACs made available by Genethon.
1994	Phred, programme for automatically interpreting sequence data, and phrap, programme for assembling sequences. Developed by Phil Green and colleagues at George Washington University.
1995	Printed glass microarray of complementary DNA (cDNA) probes utilised by Patrick Brown et al at Stamford.
1996	Affymetrix markets DNA chips.
1997	First commercial capillary sequencing machine, Molecular Dynamics' MegaBACE .
1998	PE Biosystems introduce PE PRISM capillary sequencer.

Sequencer Development (Rothman 1998)

We cannot here trace out the ramifications of all the instrumentalities highlighted by the timeline in Table 2.3. However, there is one which no discussion of the HGP could avoid, and that is sequencing. The story of the development of gene sequencers raises interesting questions about the social and political aspects of research. For example, the role of instrumentalities in scientific research, the interplay between publicly funded R&D and commercial companies with respect to both technical progress and intellectual property rights. Sequencing was first developed independently, via two distinct methods, by Fred Sanger and colleagues at Cambridge University, and Allan Maxam and Walter Gilbert at Harvard. For this achievement Sanger and Gilbert shared one half of the 1980 Nobel Prize for chemistry, remarkably this was Sanger's second Nobel. Later the UK HGP would be housed in a new centre named after Sanger. Both sequencing methods aided the development of a wide range of molecular genetic studies and complemented the other great advance of the same period, artificially recombinant DNA, for which Paul Berg received the other half of the 1980 Nobel Prize for chemistry.

Despite their elegance and efficacy the methods were time consuming and labour intensive, and throughout the early 1980s attempts were made to automate various steps in the processes, and develop what became the DNA sequencer.

The earliest successful attempt was at Leroy Hood's Caltech laboratory, where in 1986 they developed an approach that used fluorescent dyes and lasers allowing the acquisition, storage and analysis of sequence data directly by computer during gel electrophoresis. The Caltech group and a commercial company, Applied Biosystems, began a collaboration that led in 1997 to a commercial machine, the ABS 370A. A definitive history of these developments remains to be written, for at present it seems to be disputed. Some IPR aspects of the collaboration are currently being contested by MJ Research, a thermal-cycling manufacturer, which is claiming in a lawsuit against Applied Biosystems and Caltech that key elements of the Caltech work were publicly funded; work which Hood says preceded the public investment. MJ Research argue that the patents are based on work funded by the US Government and carried out by a former post-doctoral student called Henry Huang. They claim Huang thought up the idea central to the sequencer '...counting and tracking DNA molecules by tagging them with fluorescent dyes' (EN 2002). The story of the 'robbed' post-doc is a familiar claim in science; in this case Caltech state Huang left before his work could be developed, and 'what he did is not enough to rise to level of making him an inventor'. Whatever the rights and wrongs of this case no one doubts the great significance of the Caltec/Applied Biosystems work. In any case the sequencer did not long remain solely in their province. Very soon, similar sequencers were developed by other firms in the US, and elsewhere in Europe and Japan, reflecting an expanding market created in large measure by the HGP.

Even these machines were not truly fast enough to realistically sequence the human genome before the HGP completion target date of 2005 and the search was on for improvements. The DOE is reported to have spent '...over $1 billion on more than 2000 proposals from academic scientists...' (Goozner 2000) to improve sequencing speed and accuracy. In the meantime the HGP workers had to put the human genome on a slow burner and hone their skills on smaller genomes. John Sulston described their dilemma (Sulston and Ferry 2002: 72):

> It's not the difference between being able to do nothing, and suddenly having a new technique... Sequencing has been improved incrementally through machines, chemistry, enzymes and software gradually allowing us to reduce costs and automate. Like the car, though, cumulatively it's come a long way. Although we couldn't have known this in advance, the way to make progress in sequencing has been to do the best you can with the technology you have available, not to spend years trying to invent a better technology (or worse still, waiting for someone else to invent it).

The ABI sequencers were examples of such incremental improvement and over 6000 were sold over the ten years since their advent.

Several academic workers, see Table 2.3, were independently involved in a major technical advance that involved replacing the gel slabs of the original sequencers by capillary tubes. Much of this work was funded by the DOE, which by 1996 felt able to work on a prototype capillary machine – which would form the basis of the next generation of design. The DOE patents, based on publicly funded research, were taken up by commercial firms, and the first commercial capillary sequencer was marketed in 1997 by Molecular Dynamics (now part of Amersham). A year or so later Applied Biosystems followed suit with an even more powerful machine the ABI PRISM 3700. It was the entry of this machine into the actor networks of the HGP that created an enormous upheaval and made it possible for a private company to challenge the publicly funded HGP. How this was done will be dealt with in Chapter Three. Suffice it to say at this stage new technology revolutionised the HGP, causing major political problems. Ironically, in view of the later sneers about 'inefficient' public research and the greater effectiveness of private enterprise it had been publicly funded research that had laid the foundations for the new advance in sequencers. Of course, as Joseph Jaklevic, who led the DOE team who made the first prototype capillary sequencer, said one shouldn't therefore minimise the enormous effort needed to commercialise the idea. 'There's a lot of engineering that goes into making a machine that is reliable, that you can put in the box with an instructional manual and ship across the country … But once you know someone has done it, it makes it a lot easier' (Goozner 2000).

Instrumentalities and Social Control

Scientific research is not simply a game or struggle played out between scientists and nature, it is at the same time also a process of power struggles and social control between the scientists. Therefore before we close our discussion of instrumentalities we need to consider the part that they can play in the social control processes of science. In September 1989 a new strategy for genome mapping was proposed (Olson et al 1989) in which sequence-tagged sites (STSs) could be used to overcome a major strategic problem in the HGP. For David Botstein 'The strategic problem was how to connect all the physical maps together, how to get a universal language' (Cooper 1994: 123). Olson said he

> …saw large-scale physical mapping as a kind of Tower of Babel. People were subdividing the problem by chromosome and by chromosome region, and I saw us ending up with a bunch of contig maps expressed in completely incompatible languages. That is, each group was building a map from a different clone collection and was using its own method for detecting clone overlaps. Consequently we would have no convenient way to compare or crosscheck the maps … The STS idea was to annotate each contig map with a series of unique landmarks … a short stretch of DNA – between 100 and 200 base pairs long – whose base sequence is found to be unique. Since the landmark is specified by a unique sequence of base pairs it is called a sequence-tagged site … it can be unequivocally recognised and at the same time amplified using the polymerise chain reaction (PCR) … The crucial feature of STSs is that they have unique sequences. In other, words, if we determine that two clones, one from each of two different clone collections, contain the same STS, we have no doubt that they come from the same region of the genome.
>
> (Cooper 1994: 123)

STSs allowed a common language, and did away with the need for a universal central storage place for DNA clones jokingly referred to by Botstein as 'the Sears Roebuck of molecular biology' (Cooper: 124). The latter would have been enormously expensive, but, STSs allowed scientists to pick out from their own clone collection the clone containing a specified STS. Further, the STS could be digitised and transmitted by computer, allowing, argues the sociologist Stephen Hilgartner, mapping landmarks to be represented as an inscription side stepping the biological material (Hilgartner 1995: 313).

Hilgartner describes the STSs as boundary objects

> …that could tie together several diverse categories of labs – 'small' ones and 'big' ones, genetic mapping and physical mapping labs, restriction mapping labs and contig mapping labs, today's labs and tomorrow's labs, labs in the United States and labs elsewhere.
>
> (Hilgartner 1995: 308)

STSs were, in his view, a social technology as well as a biotechnology. This improved communication also opened the way towards a more centralising management style in which research centres could be made more accountable by use of metrics, e.g. accuracy, numbers STSs produced, and so forth. Hilgartner reports that this development was by no means universally popular. Another worrying issue for some researchers concerned access to scientific data and associated boundary issues since the arrival of STSs made it possible to alter data access practices, control over actual biological material was potentially no longer as important as it had been in giving someone a competitive edge. Hilgartner drew the following general conclusion from the case of STSs in the HGP, that it provides

> … a strong example of how social control in science … can be constructed: not simply through the ethos of shared norms but through technical and procedural means that build social control – directly and materially – into the very fabric of scientific production.
>
> (Hilgartner 1995: 312)

In this chapter we have concentrated on the origins and the first decade of the HGP showing how it came into being and analysing the struggle to lay the technical, scientific and organisational groundwork for achieving the ultimate goal of sequencing the human genome. In our next chapter we deal with the actual achievement of that goal, and the remarkable race to be first to sequence the genome which ensued when the public genome programme was faced with a commercial rival. There we are in a position to take our discussion of instrumentalities and social control into new and more bitterly contested territory.

Chapter 3

Doing the Human Genome Project

In the previous chapter we showed how much of the drive to establish the HGP was exterior to the field of human genetics. Certain groups, that included scientists and bureaucrats were able to mount a political lobby successfully and enlist supporters in government circles, research agencies and the media, in the teeth of opposition from many members of the biological sciences community. The scientific case for the megascience project was buttressed by organisational and national prestige, political, economic and medical health reasons. We also sought to demonstrate the importance of instrumentalities and technology in the HGP.

In this chapter we examine the evolution and development of the HGP as it progressed from launch to completion 13 years later, two years ahead of the 15 year timetable. As in the previous chapter we use timelines as historical summaries, picking out events and developments for closer analysis and to illustrate political and social issues of science.

It is convenient to periodise the history into two parts, each with its own timeline. The first period spans 1991–1997, Table 3.1, and the second 1998–2003, Table 3.2. The first period was one of steady progress in mapping, and establishing the mapping and sequencing centres, dominated by public consortia which emphasised accuracy and rapid public dissemination of data. The second period saw an intensification and speed up of the project consequent on the arrival on the scene of a privately funded rival to the Public Consortium. The ensuing struggle and competition throw up for study several important political and economic issues, such as stresses in the traditional ethos of science, and the conflict between free dissemination of scientific data and private intellectual property rights. The period ends with the establishment of genomics and several other related 'omics' emerging.

Table 3.1 Timeline Human Genome Project 1991–1997 (based on *Science* 291, 1195, 2000)

1991	
(June)	J. Craig Venter at NIH announces expressed gene strategy using ESTs. NIH files patents on partial genes.
1992	
(April)	James Watson resigns as Head of NCHGR after dispute over patenting with NIH Director Bernadine Healy.
(June)	Venter leaves NIH, sets up the Institute for Genomic Research (TIGR), not-for-profit organisation. A sister company Human Genome Sciences, headed by William Haseltine, is established to commercialise TIGR products.
(July)	The Wellcome Trust provides $95m for UK HGP.
(Oct)	Physical maps of Chromosomes Y (Whitehead Inst.) and 21 (Centre d'Etude du Polymorphisme Humain (CELPH) and Genethon).
(Dec)	NIH and DOE guidelines on sharing and releasing data. Genetic maps of mouse (Lander et al Whitehead Inst.) and human (J. Weissenbach et al (CELPH) completed.
1993	
(April)	Frances Collins appointed Director NCHGR.
(Oct)	NIH & DOE publish revised plan 1993–1998; goals include sequencing 80 Mb by end 1998 and complete human genome by 2005.
(Oct)	Sanger Centre, Hixton nr. Cambridge UK, opened. Headed by John Sulston.
1994	
(Sept)	Complete linkage map of human genome published by D. Cohen et al at Genethon and J. Murray et al at U. Iowa.
1995	
(July)	*Haemophilous influenzae* sequenced (Venter and Frazer at TIGR and Hamilton Smith of John Hopkins), first sequence of free-living organism. Physical map human genome containing 15,000 markers (Researchers at Whitehead Inst. and Genethon).
1996	
(Feb)	At Bermuda meeting, funded by the Wellcome Trust, international HGP partners agree to release sequence data into public databases within 24 hours.

(April)	NIH funds six groups to attempt large-scale sequencing of the human genome.
(Sep)	DOE initiates six pilot projects, funded at $5 million total, to sequence the ends of BAC clones.
(Oct)	Complete sequence of the yeast *S. cerevisiae* genome published by an international consortium.
(Nov)	Japanese group at RIKEN completes the first set of full-length mouse cDNAs.
1997	
(Jan)	NCHGR becomes the National Human Genome Research Institute; DOE creates the Joint Genome Institute. UNESCO adopts Universal Declaration on the Human Genome and Human Rights.
(Sep)	*E. coli* sequence, 5Mb completed by U.S. group.

One Best Way?

From the earliest discussions of the HGP there has been a powerful counter flow to the dominant strategy of seeking to sequence the whole human genome, 97 per cent of which is comprised of so-called junk DNA, masses of repetitive base sequences between genes. The argument for an alternative strategy based itself on the view that Medicine and Commerce wanted genes and therefore the initial focus of the HGP should be on finding and sequencing genes, leaving the task of a complete genome sequence to a future date. Brenner was arguing for such a strategy during the early HGP discussions of the mid-1980s. This strategy, the complementary DNA (or cDNA) sequencing approach, predated the HGP having been developed in 1983 to find muscle genes by a group at MIT (Putney 1983).

Sulston described the technique as follows:

> It involved extracting RNA from tissue and using it as a template to make…cDNA, thereby isolating the protein-coding sequences of genes. Sequencing 140–400 bases at the ends of the cDNAs gives a pair of unique tags [expressed sequence tags(ESTs)] for each. Using the ESTs to probe the sequences already deposited in the public databases, you can find out if they represent known genes, if they have similarities with genes from other species, or if they come previously unknown genes.
>
> (Sulston and Ferry 2002: 105)

A public and bitter debate followed a proposal by Craig Venter, then an NIH researcher, which presented the EST strategy as a cheaper and more efficient alternative to genomic sequencing. Venter presented his approach as a bargain by comparison with the HGP, costing a few million dollars compared to hundreds of millions. Clearly such a claim was perceived as a threat to many researchers who had nailed their research careers to the HGP mast.

Some idea of the intensity of the dispute can be obtained from Sulston's account in which he claims '... he [Venter] was clearly trying to bring about a shift in policy through his public announcements – a tactic he has employed frequently since.' (Sulston and Ferry 2002: 186) At that time Sulston was committed to sequencing the nematode worm *C. elegans* as part of the HGP and he clearly saw Venter's EST strategy as threatening. His response provides a fine example of the competitive politics of contemporary scientific research.

> We had already been personally affected by Craig's [Venter] competitive streak... we heard that his lab was sequencing worm ESTs. I don't think I'm being paranoid in suggesting that they were trying to compete with us; there has been quite a lot of publicity about our getting the sequencing grant... we held a conference call in which Craig argued forcibly that ESTs were a better way of finding genes, and he had found a gene that we had missed and so on... I saw Craig's challenge as a threat to what we were doing. If his lab was able to identify a substantial number of worm genes at a time when we only had a few dozen... it probably would not help our case for future funding.
>
> (Sulston and Ferry 2002: 107)

Sulston and his research partner, Bob Waterston, were driven by Venter's challenge to temporarily alter their research plans and sequence *C. elegans* ESTs. Sulston later wrote 'It was quite useful to have done it, but I wanted to put down a marker saying we could do this just as well as anybody else – and now we were going to get on with sequencing the genome, thank you very much' (Sulston and Ferry 2002: 107).

We should also observe Venter's intervention came during a period in which Cambridge genome workers were negotiating with the Wellcome Trust for a $95M grant, for the establishment of the Sanger Centre. Science is about political skills in garnering resources as well as technical and theoretical skills. Without necessary resources, and their continuing supply into the future, little can be achieved – consequently, we should not be surprised at the vigour with which the genome researchers defended their 'patch' and chosen approach to sequencing. Bearing this in mind we can see why funding Venter's EST project, even as a complementary activity in the HGP, was blocked.

NIH turned down Venter's EST proposal, since it was perceived as deflecting from the main goals at that time of producing the global and complete maps. Venter was forced to find his necessary resources elsewhere than in the public sector – with profound implications later for the HGP. The waters became more muddied when NIH applied to patent over 2000 partial sequences, produced in Venter's laboratory – the intellectual property rights (IPR) issue will be dealt with in Chapter 7. The row within NIH eventually led to both Venter and Watson leaving NIH. Nevertheless, although Watson lost his position as Director of the National Center for Human Genome Research, his successor Francis Collins continued his policy of going for the complete sequence. Presenting a new HGP five-year plan Collins wrote

> Although there is still debate about the need to sequence the entire genome, it is now more widely recognised that the DNA sequence will reveal a wealth of biological information that could not be obtained in other ways... Thus the Human Genome Project must continue to obtain its original goal, namely, to obtain the complete human DNA sequence.
>
> (Collins and Gallas 1993: 46)

Shotgun Sequencing, People Matter

If we examine Table 3.1 Timeline Human Genome Project 1991–1997 we can see how the EST protagonists fared in the following few years. Sulston became Director of the Wellcome-funded Sanger Centre, with its backing resources apparently guaranteed – a political job well done. Venter sought and found funds in the private sector which enabled him to create The Institute for Genomic Research (TIGR) – a not-for-profit organisation where he was able to publish as an academic scientist. The deal with his financier was that a firm named Human Genome Sciences would be allowed first look at TIGR's findings and use of its databases for commercial clients. We need not here go further into these details as they form the subject of Chapter 7. Suffice to say that TIGR made numerous contributions to genomics, the most important at this point in our discussion was the sequencing of the bacterium *Haemophilus influenzae* in 1995.

In science, significant developments can be the result of individuals of distinctive psychological types and constrasting backgrounds coming together, often by chance. An obvious case would be the meeting of James Watson and Francis Crick. Almost by chance Watson, a visiting research fellow, was allocated a lab space in the same room as Crick at Cambridge.

> Jim [Watson] and I [Crick] hit it off immediately, partly because our interests were astonishingly similar, partly, I suspect, a certain youthful arrogance, a ruthlessness, and an impatience with sloppy thinking came naturally to both of us. Jim was distinctly more outspoken than I was, but our thought processes were fairly similar. What was different was our background knowledge.
>
> (Crick 1990: 64)

Their collaboration enabled them to discover the structure of DNA. Whereas in the case of Maurice Wilkins and Rosalind Franklin at Kings College London a lack of compatibility failed to produce fruitful collaboration.

The meeting in 1993 of Hamilton Smith – Nobel Laureate for the first characterisation of a restriction enzyme – and Craig Venter is another such meeting of different minds that lead to creative advance. An account of their meeting has been provided by Davies (2000).

> As they relaxed in a bar in Bilbao, Venter and Smith became acquainted. The two men were a study in contrasts, but the shy Smith found himself liking Venter's self-confidence, while Venter was impressed with Smith's quiet intelligence and ideas. As Venter puts it, My guess is, we both wish we could be a little more like the other person.
>
> (Davies 2000: 105)

At Venter's invitation Smith joined TIGR's scientific advisory board. Smith proposed that TIGR should sequence *H. influenzae*, the organism he had originally used in his work on restriction enzymes. His method was audacious, he argued that the time consuming physical mapping phase could be omitted by using a shotgun approach. Basically this involved cutting the DNA '…into random chunks, selecting those smaller than 2,000 letters, which he [Smith] placed into a different bacterial DNA to grow and purify before running the sequencing reactions'. A special piece of software, an assembler programme, was used to piece the bits of sequenced DNA into an ordered sequence of the genome. The shotgun technique was to be TIGR's preferred method for sequencing other microbes, and several years later for the human genome. Indeed at the end of the paper that described the *H. influenzae* Venter said '…this strategy has potential to facilitate the sequencing of the human genome'. (Davies 2000: 118).

To understand the significance of this statement we need to compare it to the method then being used by the HGP sequencers. The approach that the HGP had planned for sequencing the human genome was a clone-by-clone method in which

> … the genome is chopped into chunks up to several hundred base pairs long, and cloned into bacterial artificial chromosomes (BACs). Each BAC is sequenced by shattering its cloned DNA into smaller fragments and then reassembling the sequence using computer algorithms that match the overlapping ends. Because BACs can be mapped onto the genome's chromosomes by looking for markers called sequence tagged sites (STSs), the entire sequence is gradually built up. The downside is the cost, in time and effort of constructing and sequencing in depth the library of 20,000 or more BACs that is needed to map the entire genome.
>
> (Butler 2002)

The whole genome shotgun method was being presented by Venter and Smith as a more efficient method, at least in time and cost, to the clone by clone method.

They were not the first scientists to see the potential of the method for sequencing the human genome. As early as 1993, James Weber, a molecular biologist, and Gene Myers, a software expert jointly had had the idea of using the shotgun method for the human genome. Although they were both able and creative scientists they lacked high visibility and prestige. Weber put in a research proposal to NIH but not surprisingly it was rejected. Weber presented the plan to the 1996 HGP meeting in Bermuda. This was perceived as a threat, and Weber had his ideas severely criticised by the genome sequencing elite at the meeting. The seriousness of such an attack on a relatively junior figure in scientific community should not be underestimated, and it took a further year before he and Wade were able to publish their ideas to a wider audience (Weber and Meyers 1997). Significantly their paper was not allowed to stand alone, alongside it was a critique by Phillip Green (1997), a leading bioinformatics specialist.

Green's key objections to the shotgun technique would be repeated many times during the struggle for supremacy between Venter and the Human Genome Public Consortium. He claimed that there would be thousands of gaps in the sequence, and their extent would not be clear until near the end of the project. Furthermore it was not possible to extrapolate from the shotgun method's success with bacteria, since the percentage of DNA repeats in the human was far greater making computer alignment of fragments much more difficult. His final point was most illuminating of our earlier comments regarding the politics of research.

> It is not clear how one would deal with the hiring, training, and laying off the relevant people on the massive scale required. It is also unclear how the project could be distributed among several laboratories … There is no reason to switch.
>
> (Green 1997 cited by Davies 2000: 143)

Here the balance between technical argument and politics is a fine one. Recalling his experience of the opposition from the genome elite in Bermuda, Weber later told Davies 'These large genome centres were well established to sequence cosmids and BACs, and to change to something radically different would have meant entirely overturning their labs' (Davies 2000: 142).

Some useful insights about the scientific community and creativity emerge from this case. The role of chance encounters, bringing together unlikely creative partnerships, especially if they link scientific novelty and resources. That new ideas, especially from outsiders (Wade and Myers), are perceived as a threat by the core elite group of a research network, i.e. those who have garnered the bulk of the research resource, are not welcomed, and need to be destroyed. This is contrary to the traditional idealised norm of disinterestedness whereby scientists are supposed to act without regard to careerist self interest or reputation. Finally, such pioneers, without resources to control, tend to be powerless, and their ideas once published are freely available for others to exploit.

Venter and Smith published a landmark paper in 1996 proposing a new strategy for the Human Genome Project that would make use of the shotgun technique (Venter et al 1996). Unlike Wade and Myers, they didn't need the explicit support of the human genome community. Through TIGR they controlled their own resources, and as we explain later, Venter – with new human and non-human allies – would by 1998 produce a powerful new genome sequencing network, Celera, able to rival that of the HGP. Significantly, one of their recruits would be Gene Myers.

The Strategy Changes from Steady Progress to a Race

The period of the HGP from 1990–1997, summarised in Table 3.1, showed steady progress; physical maps of the human genome had been produced, sequencing of model organisms was going well, the yeast genome completed and sequencing of *C. elegans* well underway, BAC clone ends were being sequenced, and advances in sequencing technology were in the pipeline. Chromosomes had been allocated to various groups in the HGP Public Consortium throughout the world, things were coming in place to reach the agreed goal of a complete sequence by 2005 (Burris et al 1998, Rowen et al 1997, Waterston and Sulston 1998).

An event in early 1998 threw the HGP Public Consortium's well laid plans into disarray. On 9 May, Craig Venter and PE Corporation, the world's largest sequencer manufacturer, announced that they were forming a new company that would substantially complete a sequence of the human genome in three years, and do it at a cost of $300m – a fraction of what the HGP was spending (Marsh and Pennisi 1998). The company would be headed by Venter, and called Celera. The idea for Celera seems to have come from Michael Hunkapillar of Applied Biosystems (a PE subsidiary), which manufactured DNA sequencers. Hunkapillar was one of the important pioneers of automated sequencing, who went on to found Applied Biosystems. Earlier we described its development of the ABI PRISM 3700 which revolutionised sequencing efficiency. Hunkapillar persuaded Venter to leave TIGR and to team up with Applied Biosystems in PE Corp. Both Applied Biosystems and Celera would be subsidiaries of PE Corporation. The details of Celera as a business will form part of Chapter 7.

The detailed story of how both the HGP Public Consortium and Celera completed and published their draft sequences of the human genome has been presented by Davies (2000) and Sulston and Ferry (2002), here we can only outline the key events, see Table 3.2.

Table 3.2 Timeline Human Genome Project 1998–2003 (based on *Science* 291, 1195, 2000)

1998	
(May)	PE Biosystems Inc. introduces the PE Prism 3700 capillary sequencing machine.
(May)	Venter announces a new company named Celera, which plans to sequence the human genome within three years for $300 million.
(May)	In response, the Wellcome Trust doubles its support for the HGP to $330 million, taking on responsibility for one-third of the sequencing by Sanger Centre.
(Oct)	NIH and DOE accelerate HGP with a new goal of creating a 'working draft' of the human genome by 2001, and moving the completion date for the finished draft from 2005 to 2003.
(Dec)	*C. Elegans* sequence completed by team led by Sulston (Sanger Centre) and R. Waterston (Washington University).
1999	
(Mar)	NIH again moves up the completion date for the rough draft, to spring 2000. Large-scale sequencing efforts are concentrated in centres at Whitehead, Washington University, Baylor, Sanger, and DOE's Joint Genome Institute, the so-called 'G5'.
(April)	10 companies and the Wellcome Trust launch the SNP consortium, with plans to publicly release data quarterly.
(Sep)	NIH launches a project to sequence the mouse genome, devoting $130 million over three years.
(Dec)	The first sequence of a human chromosome, number 22 completed, by British, Japanese, and U.S. researchers.
2000	
(Mar)	Celera and academic collaborators sequence *Drosophila melanogaster*, the largest genome yet sequenced, 180Mb; presented as a validation of Venter's controversial whole-genome shotgun method.
(Mar)	Disagreements over a data-release policy and plans for HGP and Celera to collaborate disintegrate amid considerable public recriminations.

(May)	Complete sequence of chromosome 21 published by HGP consortium led by German and Japanese researchers.
2000	
(June)	At a White House ceremony, HGP and Celera jointly announce working drafts of the human genome sequence, declare their feud at an end, and promise simultaneous publication.
(Oct)	DOE and MRC launch a collaborative project to sequence the genome of the puffer fish, *Fugu rubripes*, by March 2001.
(Dec)	An international consortium completes the sequencing of the first plant, *Arabidopsis thaliana*, 125 Mb. HGP and Celera's plans for joint publication in *Science* collapse; HGP sends its paper to *Nature*.
2001	
(Feb)	The International HG Sequencing Consortium publishes its working draft in *Nature* (15 February), and Celera publishes its draft in *Science* (16 February).
2003	
(April)	The International Human Genome Sequencing Consortium announce the completion of sequencing the human genome.

By joining up with ABI in Celera Venter had gained a group of strong allies in his struggle to become the most powerful actor in the human genome sequencing network. First, he obtained massive corporate financial backing, at the price of turning the sequence into a commodity – the previous perception was that it would, and should, be a public good. Second, he gained access to the best available instrumentalities, ABI PRISM 3700 sequencers, advanced Compaq super computers, and a new industrialised lab. Third, Celera would use a new technique, whole genome shotgun sequencing, which despite earlier criticism Venter believed, on the basis of successes at TIGR, was faster and cheaper. Finally, because the Public Consortium's physical maps were publicly available, they too would be recruited into his network. Thus in the space of a few months Venter established a rival sequencing network of which he was the central controlling actor. From the outset he was also to prove adept at recruiting powerful media support to counterbalance the political forces controlled by the Public Consortium. As if to underline his impatience with the 'steady' progress of the HGP, Celera's name was derived from the Latin *celeris* – 'quick, swift', with the corporate motto 'Speed matters, discovery can't wait'.

Celera's plans upset many in the human genome sequencing community. Waterston, a leading member of the Public Consortium, saw it as '…a cream-skimming approach…an attempt to short circuit the hard problems and defer them to the [public research] community at a very substantial cost' (Marshall and Pennisi 1998: 994). Other critics, harking back to Green's critique of Weber and Meyers, were sceptical of the sequence quality. Outside the Public Consortium elite there were scientists who welcomed Celera's entry, arguing even an incomplete sequence could prove useful for finding genes and advancing genetics research. Such a view, however, was contrary to the philosophy of the Public Consortium.

> Because of the intrinsic excitement of unravelling the human genetic code there is a great temptation to acquire a view of the human genome as fast as possible. New initiatives that accelerate and enhance the programme are to be welcomed and integrated into the emerging product, but none must divert us from the central aim of producing the ultimate complete reference sequence.
>
> (Waterston and Sulston 1998: 53)

The next two years prior to the announcement of the 'working drafts' of the human genome sequence were filled not only with scientific activity, but also with political manoeuvring as the two rivals each sought to strengthen their own network, and weaken the other's. The latter probably tells us more of the working of science than the former. To examine that contention we will look at several key emerging issues:

- The challenge to the Public Consortium
- Controversy over sequencing methods and access to instrumentalities
- Public research versus private research

The Challenge to the Public Consortium

By the Public Consortium is meant a network of centres and labs involved in the HGP across several countries. An idea of its composition and scale can be seen by examining the members of the International Human Genome Sequencing Consortium listed as the author of the 2001 *Nature* paper on the draft sequence (International Human Genome Sequencing Consortium 2001). It lists 20 centres, 12 in the United States, one each in the UK and France, three in Germany, two in Japan, and one in China – not all, as we shall see, made equal contributions to the sequencing. The Public Consortium had held together for almost a decade prior to Venter's challenge, holding regular meetings for assessing progress and sharing tasks, seemingly on a fairly consensual basis. Building and maintaining such a network was one of the great achievements of the HGP. Celera's arrival on the scene placed great strains on the network, since so much of its support in many countries, particularly the U.S., depended on access to Federal funding, which might be cut off if the politicians lost confidence in the leaders of the HGP.

Venter has been portrayed as a maverick, a money-seeking self-publicist, an egomaniac, and worse. Others have seen him as the model entrepreneurial scientist. Certainly he is not a typical scientist in his behaviour, thus it was no surprise when Venter presented Celera's challenge to the HGP Public Consortium through the media. This is in itself worthy of comment since it exemplifies a change in the nature of science communication. Originally professional papers in peer reviewed journals, and scientific conferences, were the usual primary source of communication to fellow scientists. Later they might be followed by a press release. A more recent development was to present a press release, generally with an embargo, immediately prior to scientific publication. In the case of Venter's challenge to scientific hegemony of the Public Consortium we simply find the press release and an interview with the distinguished science journalist Nicholas Wade in the *New York Times*. In addition there was a short briefing meeting with his U.S. HGP rivals Vamus and Collins of the NIH and Ari Patrinos of the DOE, which could be interpreted as an attempt to draw them and the resources they controlled into his own network. Yet another supposed scientific norm, that of organised scepticism was flouted, scientific knowledge claims should not be taken seriously until they had been peer reviewed and published in recognised scientific journals.

Wade's article went beyond a simple account of the Celera proposal, describing it as a 'take over of the Human Genome Project.' He raised doubts about the future of the public project, which could, he suggested, become redundant. Also, he thought maybe the American NIH and DOE programmes could integrate their efforts in some kind of collaboration with Venter. Perhaps, rubbing salt into the wound, by refocusing their effort from human to the mouse genome. How would Congress respond to this development he asked, writing 'It may not be immediately clear to members that having forfeited the grand prize of the human genome sequence, they should now be equally happy with the glory of paying for similar research on mice' (quoted by Davies 2001: 150). How should the Public Consortium respond to this threat? They could rubbish Celera's scientific and technical credentials, or rubbish Venter personally, or find a way of keeping Congress on side, or explore ways of doing a deal with Celera, or restructure their network with a new quicker strategy for getting the sequence. All were tried at various points, placing great strains upon the unity of the International HG Sequencing Consortium. A vigorous, detailed and dramatic insider's personal view of those events is provided in Sulston and Ferry.

Controversy Over Sequencing Methods and Access to Instrumentalities

Celera proposed to use the whole genome shotgun method to sequence the human genome. We have already discussed the earlier controversy in which the NIH HGP rejected TIGR's attempt to introduce it as an alternative sequencing strategy. This time the case for the shotgun was stronger. Venter and his colleagues now had several more years' experience of running high throughput sequencing labs. Celera's partner ABI made the sequencer, the PRISM 3700, which we have already described as being much more effective. Celera bought three hundred PRISM 3700s from its parent company.

To compete the Public Consortium had also to buy large numbers of the PRISM 3700, after initial fears that they might fall behind Celera in obtaining them. Such fears proved groundless. Tony White CEO of Perkin-Elmer (ABI) likened Celera's entry into human genome sequencing as having started an arms race, with Applied Biosystems as an arms dealer able to sell to both sides as everybody rushed to buy the most powerful sequencers. By the end of 1999 Applied Biosystems had sold about a thousand PRISM 3700s, at approximately $300,000 each to more than 250 labs.

Celera could afford massive computing power and the best software expertise. It also had enough funding to build state-of-the-art labs, with automation, ample supplies of technicians and reagents. By the beginning of 1999 Celera had the most advanced data centre, comparable in computing power to the U.S. defence labs at

Sandia and Lawrence Livermore. No wonder Sulston thought 'on the face of it [Celera's proposal]... looked extremely strong'.

Doubts remained, however, amongst the Public Consortium spokesmen, who reiterated their early criticisms of the shotgun approach, that it would be incomplete and full of holes. Collins was reported as saying Celera would produce the '*Mad Magazine* version' of the human genome (*USA Today*, 9 June, 1998). Sulston later wrote a more measured version of their doubts;

> ... our method, mapping first and then sequencing the mapped clones though laborious at first sight, meant you could check how good your data were as you went along. The quality of the product was guaranteed, and our priority had always been to produce a sequence that would stand for all time. There was no doubt that the whole-genome shotgun would deliver plenty of raw sequence, but how many pieces would the assembly end up in? If it turned out to be full of holes, it would have some use for gene hunting: but it would be prohibitively expensive to finish, and the location of the smaller pieces on the genome would be unknown.
>
> (Sulston and Ferry 2002: 156)

Having failed to incorporate the Public Consortium into the Celera network Venter played a clever move when he recruited an ally based in another sequencing camp. Venter needed an alternative eukaryote genome to demonstrate the efficacy of Celera's shotgun approach and found it in *Drosophila* (fruit fly). From a geneticist's point of view he could not have chosen a better organism, for historically *Drosophila* had been the site of major genetic advances (Rubin and Lewis 2000).

Since 1995 there had been an international *Drosophila* genome sequencing project, which by 1998 had sequenced about a fifth of its 120mb genome target. Venter approached Gerry Rubin, leader of the project, and offered to help complete the task of sequencing *Drosphila* in Celera. Rubin agreed, allowing Venter to test his methodology on a multicellular animal. The deal was that Celera would perform its shotgun strategy on the fly and the *Drosophila* group would contribute clone-based genome maps and low coverage sequence of the clones comprising the map.

By September 1999 sequencing of *Drosophila* was completed, in only four months (Meyers et al 2000, Pennisi 2000). Clearly a success for the shotgun method, made possible by bypassing the HGP leadership and enrolling an elite sequencing actor from a different community. The genome was made publicly available, so cleverly deflecting the criticism, widespread in the scientific community, that Celera was only interested in profit. It may well have been a 'loss-leader' but the *Drosophila* researchers, a group with the highest scientific credentials and a long tradition of path breaking advances in genetics, were delighted, thus building up Celera's scientific credibility.

In a written testimony before a Congressional Committee (Subcommittee on Energy and the Environment of the Committee on Science, U.S. House of Representatives April 2000) Rubin reviewed his experience.

> My colleagues were not enthusiastic about a collaboration with a for profit company on the genome project, despite the fact that academic researchers develop partnerships with the pharmaceutical and biotechnology industry all the time. A lot of my friends were particularly leery of a collaboration with Celera. They warned me that I was going to get

> into real trouble and would feel badly treated at the end of the day… But my interest in this collaboration was pretty simple. By combining our efforts, it seemed likely that we could get the science done better and faster than either group working alone… I am happy to be able to report that the collaboration was both highly successful and enjoyable. Celera honoured all the commitments they made to me in this collaboration and they behaved with the highest standards of integrity and scientific rigor.

Sulston, one of the most influential members of the Public Consortium, was far less generous;

> It was surprising to a number of people that Gerry [Rubin] agreed so readily. Scientists in the academic world stake claims to little plots in the field of scientific inquiry which others by and large respect. Although there are plenty of examples of 'races' to make new discoveries and arguments over priorities in the history of science, the process of grant distribution and publication makes them much less than one might expect. True if someone doesn't seem to be getting anywhere then others might move in and take over… Part of the reason why many people find Craig [Venter] hard to stomach, and why others admire him greatly, is his cavalier disregard for such academic niceties – and with Perkin-Elmer's money behind him, he could afford it.
>
> (Sulston and Ferry 2002: 158)

Public Research versus Private Research

Prior to Celera entering the field most of the funding of the HGP had been by Federal and charitable foundations. At the start of the project a figure of $3b was mooted as the requirement for the completion of the HGP, over a period 1990–2005, of course, this figure would include a far wider range of scientific work than the human genome. In the absence of any accurate global figure we can provide partial figures to give some idea of the magnitude of public funding. The U.S. genome project budget for both NIH and DOE in five year periods is shown in Table 3.3 (www.ornl.gov/hgmis/project/budget.html):

Table 3.3 US Genome Project Expenditure

PERIOD	**FUNDING ($millions)**
1988–1992	460.3
1993–1997	1091.9
1998–2002	1809.1
Total 1988–2002	**3361.3**

Actual human genome sequencing is said to account for only a fraction of this total, it is not clear what that portion is.

Much of the UK programme has been funded by the Wellcome Trust, the second largest funder after the NIH; its contributions in recent years are shown in Table 3.4.

Table 3.4 Wellcome Trust Funding to UK Genome Programme

PERIOD	FUNDING ($millions)
1998	61.273
1999	103.511
2000	121.407
Total	**286.191**

A World Survey of Genomics Research (Cook-Deegan et al 2000) has calculated the global public genomics research funding, see Table 3.5.

Table 3.5 Global Public Genomic Research Funding

PERIOD	FUNDING ($millions)
1998	721.013
1999	1141.497
2000	1805.326
Total	**3667.836**

From these figures it seems quite clear that public and charitable funding associated with the HGP has exceeded the $3 billion target, though we cannot actually say that is what the human genome sequence has cost. Further, we can see that by the time Celera entered the human genome sequence field, in 1998 that already public research in the USA alone had spent over $1555.2m (1988–1997); assuming that the rest of the world had spent over that period around a third of the U.S. total, then perhaps over two billion dollars of public funds had been spent worldwide. It is not surprising that the Public Consortium was shaken to the core by Celera's arrival on the scene. Celera claimed it could complete the sequencing of the human genome for $300m.

The media, even the serious elements, such as *Business Week*, which maybe did not truly understand the scientific and technical issues, were amazed that a newly established private company could confidently announce that it would sequence the human genome more cheaply and more quickly than the Public Consortium. Two themes would dominate the media, first the excitement of a 'race', and second, and much more important the view that somehow public research was intrinsically less efficient than private research. We will address the latter issue in this section.

Forbes (23 July 2000) wrote Celera won '… outracing a clunky government consortium.' The media had begun to present the Public Consortium as clumsy and bureaucratic, and no match for fleet footed, efficient, entrepreneurial private sector, represented by Celera. Not surprisingly the Public Consortium representatives felt hurt and misunderstood.

Members of the business community argued that the sequencing should be left now to the private sector so as to save Federal funds for other research. William Haseltine, CEO of Human Genome Sciences – which had at one time been the commercial partner of TIGR – who himself doubted the commercial value of the HGP – argued that:

> The era of government sponsored big science, in which a few laboratories receive as much as $10 million a year to analyse mostly junk DNA, while scientists doing disease related work beg for financing should end. Let private companies and charitable foundations finish the job of sequencing the human genome. National pride should come from the conquest of disease, not winning a race that was not worth winning.
>
> (*New York Times* May 21 1998, cited by Davies 2001: 152)

The primary function of a commercial organisation is to make a profit, and a profit could only be made from the sequenced genome if its contents were not freely available in the public domain. Clearly there are a number of ways in which genomes might be commodified, and detailed discussion of these will found in Chapter 7. Celera announced that it intended to seek patents on a few hundred sequenced locations thought to be of medical interest, also it would delay publishing its annotated sequenced material until paying customers had had a chance to examine it (Marshall 2000).

Celera's commercial approach was completely at odds with that of the Public Consortium which had been endorsed in February 1996 in Bermuda by 50 of its leading members representing most of the major national programmes. The so-called 'Bermuda Principles' included the following:

- All human genomic sequence data generated by centres funded for large-scale human sequencing should be freely available and in the public domain to encourage research and development and to maximise the benefit to society.
- Sequence assemblies should be released as soon as possible: in some centres, assemblies larger than 1 kb would be released automatically on a daily basis.
- Finished annotated sequences should be submitted immediately to public databases.
- These principles should apply to all human genomic sequences generated by public large-scale sequencing centres to avoid having such centres establish a privileged position in exploitation and control of human sequence information.
- To promote co-ordination, large-scale sequencing centres should inform HUGO of their intention to sequence particular regions of the human genome … (Human Genome News 1996, Marshall 2001).

Celera were seen by some leaders of the Public Consortium as bidding for the '…monopoly control to the most fundamental information about humanity, information that is – or should be – our common heritage' (Sulston and Ferry 2002: viii). Public Consortium leaders often presented a form of economic argument for retaining the Public Consortium sequencing programme; i.e. to produce a sequence which was publicly available, to all with skills and resources to access it. This was a preferable state to one which encouraged privatisation of sequence material, for privatisation of the sequenced genome could interfere with further scientific research. There was, of course, another side to the economic argument – that without adequate intellectual property rights business investment would not be forthcoming to apply genomic knowledge. In opposing Celera and other similar ventures the Public Consortium leadership opened itself up to the accusation that it was undermining the economic rationale it had originally used as part of its case for establishing the HGP.

The emergence of scientists within the HGP who sought to combine science and business was seen as threatening. An examination of John Sulston's account of his experiences in the HGP is most illuminating as to how traditional scientists felt about the clash of public and private research interests. It is widely recognised that there has been a weakening of the traditional scientific ethos in which scientists present their research publicly and freely in return for recognition of its value by their peers. A growing number of researchers now find patent and other intellectual property rights interests come before a public duty to science, and the traditional academic norms of autonomy and self-regulation are being eroded (Etkowitz et al 1998).

Sulston describes a number of occasions when he too might have succumbed to pressure to make a business deal. At one point Bridget Ogilvie and Michael Morgan [of Wellcome Trust] apparently put him under pressure to consider deals; '…I said if the data couldn't be released freely then I would resign… my view was shared by many of the governors of the Trust, and so became policy [of Wellcome]' (Sulston and Ferry 2002: 111). When Sulston speaks critically of his rival Venter it is from the position of the traditional ethos of science which he believes Venter to have transgressed. 'His whole philosophy seemed to run directly counter to everything we had fought to achieve through the Bermuda Agreements' (Sulston and Ferry 2002: 150). Sulston believed such a philosophy, apparent at the time of TIGR's deal with the firm Human Genome Sciences, impugned Venter's scientific integrity. Says Sulston 'I felt he wanted to have it both ways: to achieve recognition and acclaim from his peers for his scientific work, but also accommodate the needs of his business partners for secrecy, and to enjoy the resulting profits' (Sulston and Ferry 2002: 108). James Watson felt so bitter about Venter's ambition that he compared him to Hitler. Of course, there was more at stake than the Mertonian norms, the battle for scientific priority and recognition cannot be discounted. An outside observer might concur with what Maxine Rodinson said in another context 'War has its own laws, which apply even to the battle of ideas, and one is always led into going a little too far. It is hard to wage a polemic against someone without seeming to despise him (M. Rodinson 1977: *x*).'

There are, however, occasions when public and private research entities may find it convenient to collaborate and make public their research findings. In 1994 Merck built a non-profit EST database, the Merck Gene Index. What seemed an altruistic act was also a calculated business ploy, a defensive tactic to counter for-profit databases being created and marketed by gene-based companies such as Human Genome Sciences and Incyte Pharmaceuticals. A similar stratagem occurred in April

1999 with the establishment of the SNP Consortium, whose members included the Wellcome Trust and 10 large pharmaceutical companies. Its task was to find and map 300,000 common DNA sequence variations, single-nucleotide polymorphisms (SNPs) in the human genome. The map was made public for use by biomedical researchers. This is an example of pre-competitive research, and firms were happy to sign up to because they believed it helped to prevent privatisation of SNPs during a period when it wasn't clear what their value would be (Marshall 1999).

Reorganising Networks

During 1998 it became apparent that the Public Consortium needed to increase the speed of its work to meet the challenge of Celera. To do this it would be necessary to increase funding, change the strategy and reorganise the sequencing work.

The Wellcome Trust agreed in May 1998 to double its support for the Sanger Centre to $330m, which would provide Sanger with sufficient resources to sequence a third of the genome. The continuing and increased support by the Wellcome Trust seems to have been of vital importance to the whole consortium. As a charitable foundation it was independent of government and shielded from the political pressures that were plaguing the U.S. HGP. The Wellcome Trust's decision was called ' a shot in the arm' by Collins, and Watson claimed 'it was absolutely critical, psychologically (Sulston and Ferry 2002: 167).' It was said to have stiffened the resolve of the Americans to resist demands for them to collaborate with, or even make way for, Celera.

By October 1998 the NIH and DOE had agreed to a new strategy which would accelerate their programme at the cost of lowering the sequencing standard. A 'rough or working draft' would be produced by the end of 2001, this target was later brought forward to spring 2000. The rough draft would '...comprise shot gun sequence data from mapped clones with gaps and ambiguities unresolved' (*Human Genome News* 1999, Pennisi 1999). The date for a 'completed high quality human DNA reference sequence' would be sometime in 2003, the 50th anniversary of the elucidation of the structure of DNA by Watson and Crick. The Public Consortium had been forced by Celera's competitive pressure to succumb to what was an intermediate product. The speed-up led to a reorganisation and concentration of the sequencing effort into five principal centres, the so-called 'G5', whilst other less well endowed centres in the USA, Japan, Germany and France were allowed smaller parts of the genome (See Table 3.6).

Table 3.6 Chromosome Assignments for the Public HGP Centres (Davies: 164)

Laboratory	Director	Chromosomes
*Sanger Center	John Sulston	1,6,9,10,13,20,22,X
*Washington University	Bob Waterston	2,3,7,11,15,18,Y
*Baylor Coll. of Medicine	Richard Gibbs	3,12,X

*Joint Genome Inst.(DOE)	Elbert Branscomb	5,6,19
*Whitehead Inst./MIT	Eric Lander	17 [+others]
University of Washington	Maynard Olsen	7
Genome Therapeutics	David Smith	10
France	Jean Weissenbach	14
Germany	Andre Rosenthal, Helmut Bloecker, Hans Lehrach	8,21
Japan	Yoshiyuki Sakaki Nobuyoshi Shimizu	8, 18, 21, 22

*Member 'G5' group

Such a reorganisation came at a price, and for all their angst about the traditional ethos of science, sharing chromosomes, international collaboration, and so forth the leaders of the Public Consortium showed that they were capable of being quite as ruthless as their commercial rival. In this context the thoughts of Sulston, a major actor in the reorganisation, are again illuminating. 'The very existence of the G5 was slap in the face to colleagues who had participated in the Bermuda meetings since 1996 and regarded themselves as partners in the consortium' (Sulston and Ferry 2002: 191). A two-tier structure had been created favouring the most powerful group – the G5 – at the expense of the others, a classical example of the Mathew effect in the allocation of resources (Merton 1973: 457). The group dynamics were affected, especially the internationalism of the community; the other international centres outside the G5 believed that they had lost out. The French, Germans and Japanese had a role says Sulston '… but only as long as they could keep up with the rest … they were on board, but there was no doubt at all who was in the driving seat' (Sulston and Ferry 2002: 194–195).

Even within the G5 technical and political problems associated with the speed-up of the programme threatened what Sulston termed the 'fragile consensus'. On the technical side, there was a problem with the supply of BAC clones necessary for the sequencing work. Some groups such as the Sanger had access to the necessary BACs, others, in particular the Whitehead Institute group headed by Eric Lander did not have enough of the mapped clones they needed to fulfil their sequencing goals. The latter then argued that maybe they should be allowed to sequence BAC clones at random from a library covering the whole genome.

The random approach, however, had an important political implication. It threatened the agreed 'regional' strategy whereby consortium members worked on allocated chromosomes or regions of the genome. Technically the regional approach allowed better quality control. Significantly, according to Sulston, this approach also aided the politics of the Public Consortium by allowing work to be genuinely collaborative, sharing work according to research capacity and interest. The random approach on the other hand discouraged such collaboration since it might encourage 'cherry picking' commercially important genes. There was a period when the British team at Sanger and their supporters at the Wellcome Trust perceived that the Americans were moving towards a 'hybrid' strategy, some groups working regionally and others randomly, and that this might be the first of moves towards a compromise with Celera. Eventually the BAC supply was resolved for Lander's Centre, allowing the survival of the regional approach.

Some idea of the tensions produced from that debate, and centripetal potential within the consortium emerges from Sulston's response to hearing that the Americans were seriously considering adopting the hybrid approach. He likened his position, as Head of the Sanger team, in an e-mail to Francis Collins, to a British tank commander in the First Gulf War coming under 'friendly fire'. He saw the developing position as a political strategic U-turn designed to strengthen the American groups at the expense of the international partners of the consortium. 'I felt strongly to relinquish 'our' chromosomes would greatly diminish our influence on the project as a whole' (Sulston and Ferry 2002: 174).

Personal rivalries were also important judging from Sulston's observations on his American G5 colleague Eric Lander whom he perceived to be overly ambitious and seeking leadership of the sequencing programme. Why should this have mattered, was he a personal rival for fame, or was his ambition a threat to the group harmony? Sulston, perhaps a little ingenuously, tends to present himself as ambitious only for the group – or perhaps for the publicly available sequenced genome in itself. 'I never wanted to get involved in the three ring circus of the Human Genome Project … what I wanted to do was to read the genetic code of the nematode worm. I didn't imagine that the worm was going to lead us directly to the human genome' (Sulston and Ferry 2002: 5).

Sulston was not, however, averse to sacrificing small sequencing groups during the restructuring phase. What the G5 now reflected was the interests of the large industrialised laboratory, with hundreds of employees, and run like a business; the traditional lab with independent scientists and a few technicians had been superseded. Clearly the managerial responsibilities of lab leaders like Sulston and Lander had qualitatively changed, and consequently so had their behaviour and attitudes. They became convinced that the original modus operandi of the Public Consortium would not be efficient enough to meet the challenge posed by Celera without centralising sequencing, strategically it was the only way in which they might prevent the human genome being privatised. That however was cold comfort to their colleagues in smaller groups who had been effectively sidelined.

Races in Science: Winners and Losers

By the end of 1998 all parties were geared up to maximise sequencing. The popular view, in the media, was that a race had begun. Before the race with Celera the Public Consortium had the goal of a completed DNA sequence of the human genome with a 99.99% accuracy. To have a chance of beating Celera the finishing line had to be moved, becoming the working or rough draft, with lower accuracy and holes. After that the Public Consortium would continue to its original line, arriving there in 2003, two years earlier than had originally been planned.

What are races in science and technology, what do we know of them? Races in military technology have been common place in the 20th Century. The atomic bomb, the hydrogen bomb, and the space race spring to mind. In chemistry there were races to identify or create new elements, and in physics races to find theoretically predicted particles.

Molecular biology too has its races for priority. James Watson's account of his part in the race to discover the structure of DNA before Linus Pauling has entered scientific folklore, although in this case Pauling did not know that there was a race. One might also cite the race to be the first to synthesise a human gene (Hall 1988); and races to isolate specific cancer genes like BRCA1 and 2. Sulston has described how he was concerned he might have to race Venter for the worm sequence; as it happens the feared race was averted and Sulston and Waterston were able to announce their worm sequence in 1998. Nevertheless Sulston's ruminations on the potential race are worth reproducing here:

> Why did it matter to be first? It's not just a matter of personal pride. Our labs by this time were large-scale enterprises, employing a lot of people and with a great deal of hard won money invested in them. We had set ourselves tough goals and to meet them in the face of competition in order to justify the investment that had been made in us, and keep the credibility that would ensure our further funding. These are the forces that push science forward efficiently.
>
> (Sulston and Ferry 2002: 159)

It is not really surprising that scientists find themselves in situations that lead to races. After all in any field at any given moment there are only a limited number of key or important problems or research tasks, often fewer than the available talented people and groups seeking to solve them. The ability to function as a scientist depends on access to resources, both financial and intellectual. These are easier to obtain if one is highly regarded by one's peers and research awards agencies. Being first to make a major discovery provides priority over potential rivals both for prestige and intellectual property rights. Winning the race provides prestige, which in the scientific community is an important component of social power. However, it is not always clear who has won the race, and priority disputes have long been an integral part of the development of science, and an important topic of study by sociologists of science. Robert Merton described how science as an institution

> …incorporates potentially incompatible values: among them, the value of originality, which scientists to want their priority to be recognised, and the value of humility, which leads them to insist how little they have been able to accomplish… [this] generates a distinct ambivalence towards claiming of priorities.
>
> (Merton 1973: 305)

This ambivalence may take the form of denials that there is a priority dispute or race. Thus we find Francis Collins telling a Congressional Committee '…the private and public genome sequencing efforts should not be seen as engaged in a race'. Despite the fact that Celera's ambition to sequence the human genome had caused a massive upheaval in the Public Consortium.

Another form of response by the Public Consortium was in the area of media relations, the media were used to hype any achievement in an endeavour to counter the powerful public relations machine of its rival. Prior to the announcement of the draft sequence there are several landmark events that might tell us something about the struggle to get media coverage in favour of the Public Consortium's achievements. The billionth base entered the GenBank on 17 November 1999, this was a cause for celebration and media attention. Awards were given to G5 leaders by ministers in UK and U.S. The U.S. Secretary of State for Health and Human Services claimed 'The 21st Century came about six weeks early (Davies: 193).' The publication of the sequence of chromosome 22, the first chromosome sequence, on 2 December 1999 (Dunham et al 1999) was again the occasion for much hype. First it was presented as a vindication of the Public Consortium sequencing approach. Then came some eyebrow-raising comparisons, especially when it is recalled that actually a small part of the chromosome wouldn't sequence. 'As important an accomplishment as discovering that the earth goes round the sun, or that we are descended from apes' (John Sulston). 'Comparable in achievement to the invention of the wheel.' (Michael Dexter, Wellcome Trust) 'A new era has dawned, we have fulfilled the dreams of Mendel…et al' (Bruce Roe). 'Like seeing the surface… of a new planet' (Mark Pattersson, *Nature*).

As we have already said, the obvious reason for all this hype was the Public Consortium leadership's perception that Celera had previously seized the public relations initiative. It was engaged in the modern politics of science; even in the chromosome paper there is space for a political jibe at Celera: 'Over the course of the project, the emerging sequence has been made available in advance of its final completion through the internet sites of the consortium groups.' In a *Nature* article describing the completion of the Chromosome 22 sequence a Public Consortium researcher makes an adverse reference to Celera: 'The self-sufficiency of Celera's whole-genome strategy will never be put to the test… Celera's proposed sequencing of the human genome will fully exploit and be utterly dependent on publicly available HGP mapping and sequencing data (Butler 1999: 448).'

The 'real world' of business took note of the Chromosome 22 sequencing and the accompanying media hype, and there was an increase in speculative investment for genomics and biotechnology firms. Ironically one of the main beneficiaries was Celera (see Chapter 7 table 7.8).

President Bill Clinton and Prime Minister Tony Blair were also drawn into the politics of the HGP, and with an unexpected result. A statement on the HGP released on 14 March 2000 by Clinton and Blair included the phrase, 'To realise the full promise of this research, raw fundamental data on the human genome, including the human genome and its variations, should be made freely available to scientists everywhere'. Unfortunately this reversed the effect of the chromosome 22 hype on the 'real world' and caused a crash in biotech shares. This is one of the more peculiar events in the development of the HGP. Davies suggests that it might have had something to do with a breakdown in collaboration discussions between Celera and NIH; we have already seen the possibility of such collaboration caused Sulston and the Wellcome Trust much anxiety. On 5 March 2000, the week before the Clinton-Blair statement the Wellcome Trust published a letter that NHGRI had sent to Celera. The letter listed fundamental differences relating to Celera's insistence on controlling access to its sequence data, and that consequently NIH was breaking off attempts at collaboration with Celera (Marshall 2000). According to Sulston the meeting between Clinton and Blair resulted from lobbying by Mike Dexter of the Wellcome Trust; the juxtaposition of the leak and the meeting between Clinton and Blair was a coincidence. Whatever the actual reason it conveniently succeeded in placing the patenting of the human genome in the public eye.

During the first few months of 2000 there was a period of claim and counter claim about who was in the lead in sequencing. Through a series of press announcements Celera won a hype war and the press seemed to believe that the Public Consortium had fallen behind. The Public Consortium, particularly the British members, who had, for reasons already mentioned, greater freedom of action responded vigorously. The competition provoked strong feelings. 'The gloves really came off…' says Sulston, 'I pointed out that our problem was that Celera not only collected their own data but would "hoover up all of ours" – which of course, was publicly available – call it their own, and charge others for using it. "It's a sort of con job…" I had entered the world of politics' (Sulston and Ferry 2002: 218–219).

Entering 'the world of politics' clearly opens up the possibility that the politicians will seek a direct hand in matters. We have discussed the Clinton-Blair statement, and it became apparent that Clinton wanted to take further interest in the progress of the HGP. Seemingly he became impatient with the public infighting between Celera and the Public Consortium. He perceived that it might be detrimental to Vice President Al Gore's presidential campaign, for Republicans viewed Celera favourably. Clinton is said to have ordered the NIH and DOE genome leaders to come to an arrangement with Venter. The arrangement was that there would be a joint announcement of the completion of the draft sequence, joint publication and a truce in the media war – both teams would win. Additionally, Lord Sainsbury – UK Minister for Science – went to the White House and arranged that there would be a simultaneous USA-UK announcement on 26 June, apparently the only date that both politicians had free (Sulston and Ferry 2002: 224).

Clinton and Blair wanted to bathe in reflected glory; on 26 June 2000 Bill Clinton in Washington with Craig Venter and Francis Collins at his side, and simultaneously in London Tony Blair, the Sanger and Wellcome, announced the completion of the two versions of draft sequence. Officially the result had been declared a draw. Elsewhere we have referred to the hyperbole and hype the achievement generated. But since the completion date was totally arbitrary how much of the human genome sequence had actually been produced on the day? Sulston is very revealing about this:

> It was not clear that the Human Genome Project had quite got to its magic 90 per cent mark by then, and Celera's data were invisible but known to be thin, so nobody was really ready to announce; but it became politically inescapable to do so. We just put together what we did have and wrapped it up in a nice way, and said it was done. We were sucked into doing exactly what Celera has always done, which is to talk up the result and watch the reports come out saying it's all done. Yes, we were just a bunch of phoneys! But we were trapped by Washington politics'
>
> (Sulston and Ferry 2002: 224).

By 26 June 2000 there had only been press releases and public display, the publication of the two working drafts remained to be done. This in itself is further evidence of just how much traditional values of scientific probity had been undermined – though to be fair the Public Consortium, unlike Celera, had regularly placed its completed sequences in a public database, GenBank. Nevertheless research is not completed until placed in a peer-reviewed journal. A published peer reviewed paper would remove some of the chicanery of 26 June. It had been intended that both Celera and the Public Consortium would publish their work as a special journal issue, probably with *Science*. In the event it did not prove possible to publish both papers in the same journal, and the ensuing argument became the basis for more public quarrelling between the two teams. Furthermore, the situation that ensued raises some important questions about commercial interests and acceptable journal publication policy. The leading scientific journals fulfil an important role in publishing research, but they are also businesses. Those journals with the highest citation impacts and largest circulations are usually the most profitable for their publishers. Their editors often compete to publish research expected to have the most scientific and public impact. The two most prestigious journals *Nature* and *Science* were in competition to publish the human genome papers.

It is not possible, or desirable, to publish long complete sequences in a journal, therefore molecular biology developed publication norms to deal with this. Journals publish papers without the sequences as long as the sequences are deposited in a public database such as GenBank. This allows other scientists the opportunity of checking and criticising, if necessary, the sequence. Peer reviewed publication, in spite of its known drawbacks, remains the bedrock of modern science's credibility and utility. Celera was not prepared to follow this practice, claiming that placing the sequences in its own commercial database amounted to the same as publication in a public database. Not surprisingly scientists in the Public Consortium disagreed that this would be acceptable, believing that Celera was undermining the normal peer review process accepted in molecular biology.

The editor of *Science* acquiesced to Celera's demand. As a result he received many protests about the terms of publication that he had agreed with Celera. One email from leading academics expressed concern that this set a precedent that might 'open the door to similar withholding of information by future authors, with unfortunate consequences...' (Marshall 2001: 1191). The 'unfortunate consequences' presumably refers to loss of quality control, and brings to our attention Ravetz's view that 'The problem of quality control in science is ... at the centre of the problems of the industrialised science of the present period' (Ravetz 1971: 288).

Science published a special announcement on 6 December 2000 stating that Celera had agreed to a plan which ensured free access to its database for academic researchers, whilst charging commercial users. *Science* argued that its agreement with Celera did not breach its standing policy of requiring authors to deposit relevant results and archival data in publicly accessible databases. The next day the Public Consortium leadership clearly unsatisfied with *Science*'s editorial decision decided to submit their publication to *Nature*. In February the two groups published in their separate journals, Celera in *Science* (Venter et al 2001) and the Public Consortium in *Nature* (International Human Genome Consortium 2001). The difference in style appears even in the two titles, the Public Consortium speaks of 'Initial sequencing', whilst Venter's says 'The sequence'; as an addition piece of chutzpah one notes that the *Nature* issue carrying the Public Consortium papers also contained a prominent advertisement for Celera claiming: 'The most complete assembly of the human genome... Celera's ordered and oriented Human Genome is the most accurate and complete.'

Draft did not mean done. It was not until 14 April 2003 that the International Human Genome Sequencing Consortium announced that the sequence was done, well almost. According to a report in *Science*

> ... the new product is much more complete and of higher quality: 99% of what can be done with current technology is now done... virtually all the bases are now identified in their proper order, which was not true of the draft versions... The consortium set out to meet a standard of one error in 10,000 bases, the new version is 10 times better than that.
> (*Science* 2002: 409)

The announcement was timed to coincide with the fiftieth anniversary of the publication of the structure of DNA. Had overtime been worked in the five great sequencing centres to ensure completion by this iconic date? On this occasion, surely one of a greater achievement than that of the draft sequence, public and media interest was less, and the American President and British Prime Minister were absent. The Public Consortium had delivered the sequence, or most of it, two years ahead of schedule, at a cost of $2.7 billion. Its achievement is admirable and now we are faced by a realisation of just how much remains to be understood. To take what might seem a very basic question, how many genes do humans have? At the start of the HGP people thought there might be 100,000; then according to estimates based on the two draft sequences there were 35–45,000 genes, and at the time of writing the estimate is just over 24,000 (Pearson 2003: 576).

Conclusions

The community of biological sciences has changed much in the last 50 years, the experience of the HGP mirrors most of these changes (Collins et al 2003). The most obvious change is that of scale, contemporary research demands greater expenses, and ever more instrumentalities – of which the human genome sequence is one. Another key change is in the importance of technology and instrumentalities, which have become evermore integral to contemporary biology, thus one finds debates about the best technique to use can be as vigorous as any debate over theory. Other necessary changes have occurred in the management of research. The expenses and resources now involved in advancing a field call for a new level of strategic management of research, and focussing and concentrating research resources into fewer sites. Research leaders need to be able to manage large numbers of specialist researchers, with different skills, and keep them working as a team. Political skills, as well scientific ones, are required for the internal management of diffused megaprojects like the HGP, where participating institutions and labs are spread over many countries, and where a combination of top-down and bottom-up management applies. Further, and most importantly, such leaders need the political *nous* to ensure continuity of programmes over many years by keeping government officials and politicians on board.

The political skills of many of the HGP leaders are now being applied to redeploy the institutional and bureaucratic framework created during the HGP. For, the question arises as to where the HGP goes from here. Predictably we shall witness that other characteristic of megascience institutions, which is they rarely disappear and disband once they achieve the original goal. Goalposts will be repositioned, and the original vision replaced by new, and hopefully, grander visions. One expects that the availability of the human genome sequence will open up new scientific fields. Genomics obviously, but other 'omics' are being developed or mooted, these include functional and structural genomics, proteomics, and pharmocogenomics, we explore some these developments in Chapter 8.

Another change, over the last 25 years, which is commonly associated with the emergence of biotechnology, is the closer association of academic and commercial interests. The HGP experience demonstrated how great are the concerns and problems, as well as opportunities, resulting from the actions of scientific entrepreneurs able to move between academic and commercial research institutions. Fears have been voiced as to their effect on the traditional public ethos of science, and for quality control in science. Finally we have hinted at the commercial potential of research findings in genomics – how and where can research findings in the new genetics and genomics be effectively applied? These issues will form the subject of Chapter 7.

Finally it is, in a sense, ironic that the 'finish' in 2003 should have coincided with the DNA double helix fiftieth anniversary – an archetypal piece of individualistic creative small science. The mythical stories of such inspired individualistic discovery perpetuate misunderstanding about contemporary biology; creative genius has, as our account has shown, a place in the megaproject but it is a subordinate one under the control of whoever dominates the project's actor network. After all, Brenner had identified the HGP as 'a bit like Stalinist Russia' (Davies 2001: 152). Not everybody thinks it wise to allow a decline in small-science research labs. It may, perhaps, prove possible to bridge the gap between the needs of small-science researchers and those in megaprojects. For example, programmes are emerging in the U.S., in the light of the HGP experiences, for a third-way model of biological research (Frazier et al 2003). Large scale industrialised laboratory and database 'core-use facilities' are planned, in fields such as proteomics, which will be made available for use by small-science researchers, whose resourcefulness and energy might otherwise be missed by over-emphasising the big-science model.

Chapter 4

Managing Genetic Information

In this chapter we discuss three related issues involved in managing different kinds of genetic information: biobanking; privacy and personal genetic data; and bioinformatics. Managing genetic information in society normally occurs through a combination of professional ethical codes, and state regulatory regimes. Personal genetic information is usually defined as including: genotype – detail at the level of DNA or protein; phenotype – observable outcomes in terms of physical characteristics; and family information – recorded patterns of inheritance of different phenotypes. Genotype information is generally obtained either by direct analysis of DNA or of proteins or other bodily chemicals. Phenotype information is gathered in many different ways including tests of tissue or fluid samples, X-rays, or physical observation and measurement. Family information is normally gathered through family histories taken by clinicians and health professionals. The personal or social significance of these different forms of information varies significantly. The other main type of genetic information is the data generated through the mapping and sequencing of genomes, particularly the human genome, and stored in digital form in public databases. Attempts to deal with issues associated with managing the data explosion generated by the HGP and its applications include scientists working collaboratively using electronic forms of communication.

Biobanking

The completion of the mapping of the human genome has generated a vast quantity of raw sequence data that is of little use until the function of a particular gene and its role in pathology is known. The move to functional genomics discussed in chapter three and facilitated by the developments in bioinformatics has created an opportunity to co-ordinate genetic samples, and medical and genealogical records, into databases for the development of linkage studies, association studies, and pharmacogenetics (Martin 2001). New fields of research such as genetic and molecular epidemiology are emerging, aiming to systematically apply traditional epidemiological methods to the study of environment-gene interaction. Such studies, however, are also dependent on access to large population-based sample collections and patient records. A series of such public and private sector genetic databases have been developed in the UK over the years (Martin 2001: 166–167) using the National Health Service as their major research resource. Such developments have significant social implications, and some of these are explored in more detail below.

A controversial attempt by deCode Genetics and Swiss-based pharmaceutical giant Hoffman-LaRoche to develop a Health Sector Database began in Iceland in late 1998. It was to comprise of a computerised genealogical database of 600,000 living and dead individuals and their family relationships, and genetic sequences derived from blood samples collected by local doctors from the whole population. Iceland had been chosen in part because of its small (some 280,000 people) and apparently homogeneous population, and because of its detailed historical kinship records handed down over many generations – a hangover from its exotic Viking past. While some public debate had occurred in the Icelandic Parliament prior to its establishment, with promises made about improving not only the health but also the wealth of the Icelanders, many of the more troubling social, legal and ethical implications were never fully discussed. In particular issues about which Icelanders could 'opt out' of the Database, and on what grounds, were left vague until after the process of tissue collection had begun (Sigurdsson 2001, Rose 2001). Many Icelanders, with little understanding of international capital flows and the vagaries of the stockmarket, who had invested their life-savings in the enterprise, were left to face financial ruin when the NASDAQ plummeted following the Clinton-Blair statement on gene patenting and ownership in March 2000 (Fortun 2001).

Estonia, with a population of 1.4 million people, decided to construct a comprehensive genetic database at about the same time as Iceland (Sigurdsson 2001). The Human Gene Research Act was passed by the Estonian Parliament in December 2000 with the expectation that up to 1 million Estonians will ultimately deposit their genotype. There appeared to be little by way of opposition or critical discussion to the establishment of the database, which requires informed consent for a deposit to be made. Unlike Iceland, where issues of privacy and the commercialisation of knowledge were key issues in the debate, there is evidence that the Soviet past has caught up with Estonians who are more concerned that any data kept on them might be used by blackmailers against them in the future (Sigurdsson 2001:109).

The UK Biobank is a more recent development to facilitate exploration of the roles of nature and nurture in health and disease. Established in 2003 with funding from the Wellcome Trust, the Medical Research Council and the Department of Health, it involves some 500,000 volunteers aged 45 to 69 who complete lifestyle questionnaires and provide a blood sample for DNA and other analysis. Together with their medical histories, this information is anonymised to create a national database – the UK Biobank. It aims to provide a resource for research into some of the common disorders of later life including heart disease, cancer and type 2 diabetes. The Bank is organised on a 'hub and spoke' model, with the Co-ordinating Centre located in the University of Manchester, with responsibility for data management, quality assurance, computing and financial management, The six scientific Collaborating Centres, made up of a network of academic and research institutions, are responsible for participant recruitment and initial data and sample collection.

These developments give rise to a number of ethical and policy issues and concerns including questions about how information is obtained (Martin and Kaye 2000), and how it is used both within the public health system, and also outside in the commercial environment (Martin 2001). Information and tissue are required from many hundreds of thousands of individuals if any biobank is to become a useful tool. Pilnick (2002: 119) suggests that it is difficult to arrive at a meaningful notion of informed consent in seeking agreement to participate, particularly if the future uses of the donations are not clearly specified. Consent is not just about evaluating a particular technology, but also about assessments of trust and risk in professionals and institutions. Rose (2001: 124) notes that the scandal in Alder Hey Children's Hospital and the Bristol Royal Infirmary in the UK, surrounding the collecting of body parts, may have contributed to undermining public trust in both. Such concerns may be magnified when commercial organisations become involved as in Iceland. She also notes that medical records already in existence have often proved to be inaccurate or misleading, and no steps appear to be being taken to ameliorate the situation.

A related difficulty can be identified in the context of recording family histories that are an integral part of developing the biobanks (Pilnick 2002: 96–97). Any genetic disease or disorder must effect families as well as individuals. However, research has shown that understandings of inheritance, family or kinship are culturally structured and not necessarily held in common, even within kin networks (Parsons and Atkinson 1992). Richards (1996) argues that this can become problematic when geneticists attempt to infer biological relations from kin relations. Not only are families often not defined solely by blood relations, but where families have split, members may be difficult to trace or such tracing may be seen as problematic. This makes reliance on family history as an aid either to genetic diagnosis or to provide the basis for comparative data collection in a bank extremely problematic. The further issue of whether information about family members transmitted to a third party infringes their individual rights will be discussed later in this chapter.

The introduction of the new biobanks and genetic databases in Iceland and Britain has resulted in a need to rethink how they are best regulated. In Iceland, the hybrid nature of the proposed Health Sector Database, incorporating both the government and the privately owned deCODE Genetics, masked the complexities that underpinned the debate about regulation. The process took place in an atmosphere 'which might be described as a mixture of a shouting match, elaborate advertising campaign by deCODE (not merely innocuous town meetings), overwhelming media attention and marathon sessions in the Icelandic Parliament' (Sigurdsson 2001: 110). This atmosphere contributed to the passing of the Biobanks Act in May 2000 which further entrenched the acceptability of letting presumed consent pass for informed consent in biomedical research.

The system of governance in the UK, on the other hand, historically depends in large measure on the norms of professional conduct instilled as part of their training in scientists and clinicians. The combination of the erosion of public confidence in the medical profession, noted by Rose, and the emergence of a growing market for commercially valuable genetic data, has thrown the adequacy of current arrangements into question (Martin 2001). In contrast to events in Iceland, there has been a concerted effort to involve the public in informed discussion about the key issues. The UK Human Genetics Commission stepped in to provide informed advice to government through a consultation paper, *'Whose hands on your genes'* in November 2000 (HGC 2000). It also commissioned MORI to conduct a survey of a sample of 1000 derived from a People's Panel, including two 'boosters' from ethnic minority groups, and held two public consultation meetings in Newcastle. Together, the results of the consultation were produced as a report *'Inside Information'* in May 2002 (HGC 2002). There is little evidence as yet to suggest that these efforts have been successful in generating informed debate among the public at large, for reasons discussed in greater detail in Chapter 8.

Privacy and Personal Genetic Data

Much of the information yielded about individuals through the new genetic technologies will be 'morally inert', but some will have predictive value in a number of areas including for example, general health or life expectancy. The accuracy will vary, but the majority will be statistical and relatively indeterminate in nature. However, a number of ethical issues, with complex social consequences, still arise. The two main ethical principles are autonomy (self-determination) and privacy (control), and a balance needs to be struck in the interests of third parties and the public interest (Wood-Harper and Harris 1996). The rapid and exponential growth of available genetic information requires new forms of governance and regulation that recognise the primacy of the individual wherever possible.

Genetic material can be stored in a variety of ways that define the link to their origins (HGC 2002). Simple linkage allows easy identification of the individual donor or health record. Reversibly anonymised linkage is based on a process of encoding the identifiers so that no obvious source can be identified. Irreversibly anonymised linkage, as its name suggests, makes any future identification impossible. In some circumstances this latter form of linkage may not be meaningful in practice as, for example, in the case of some rare genetic diseases. In fact the very nature of DNA will always limit complete anonymisation since an individual's unique identity is embedded within it. However, the degree of security to be assured through anonymisation will be an important factor in obtaining informed consent.

A significant factor lies in the use to which any anonymised data might be put in the longer term, and the control that can be exercised by an individual over that use. This may be important where the medical research undertaken is of a fundamentally different nature to that for which the original consent was given. This may not be seen as problematic within the confines of medical research, where confidentiality rather than anonymity are the norm. However, the Health and Social Care Act 2001 entitles the Secretary of State to authorise the use of patient information in research without seeking patient consent to this use. This rule of confidentiality can be breached under the provisions of the Health and Social Care Act only when seeking consent is impracticable and it is not possible to use anonymised data, as for example is the case with cancer registries. Here the importance of the public good is deemed to outweigh the rights of the individual.

However, the boundaries begin to blur when large-scale data banks appear, as already illustrated in the cases of Iceland and Estonia. In particular the issue of obligation on the part of the users, and their understandings of the nature of the social contract that participation involves, come into question. In Iceland, suggest observers (Fortun 2001, Sigurdsson 2001, Rose 2001), these issues are unclear as deCODE is a commercial organisation and not a public health laboratory. The concept of obligation underpinning the prevailing view within medicine in the UK derives from the classic analysis of blood donation by Titmus (1970). He identified the values and social relations of altruism and collectivity that formed the foundations of a more efficient blood transfusion service than the one prevailing under the commercial and individualised system found in the USA. The expectation by the blood donor is that, should it become necessary, access to blood similarly donated by others will be freely given. It is unclear whether the Icelandic people will have similar access to all of the products that will emanate from deCODE in the future.

In the UK concerns about the future use of data from the UK Biobank have been raised in part as a result of events in Iceland (Martin and Kaye 2000). One concern is located in the changing nature of the doctor-patient relationship in the face of commercialisation and the development of new patent regimes (Spallone and Wilkie 2000) which is discussed further in Chapter 7. While the Human Genetics Commission is relatively sanguine in its conclusion that 'the question of commercial exploitation of commercial involvement in research or access to genetic databases should be fully explained at the time of obtaining the participants' consent' (HGC 2002: 97), it overlooks an earlier discussion concerning the process of obtaining consent. Here HGC admits that there is no clear legal authority on the issue of the amount of information that must be imparted before consent (for a genetic test to reveal genetic information for example) is given. It concludes that 'the nature and extent of information that is required in seeking consent for a genetic test depend on whether the test in question is likely to reveal sensitive genetic information' (HGC 2002: 51). In the case of commercial exploitation of data held in large-scale databases the down-stream therapeutic applications are still to be developed, and their potential for encroaching on an individual's rights unknown. This makes 'informed' consent an almost unachievable goal even if the health professional involved felt that it was relevant.

Insurance and Employment

A concern relates to the use of genetic information outside the clinical context particularly in insurance and employment. Under the UK Race Relations Act 1976 (as amended) it is illegal to discriminate on grounds of ethnic origin in the processing of information by employers or insurers regardless of consent. In contrast, the Sex Discrimination Act and the Disability Discrimination Act allow discrimination in insurance ratings on the grounds of gender or disability, providing these ratings are backed by appropriate actuarial data. Some recourse in cases of discrimination may be had under the Equal Opportunities Act 1976. In October 2001 the UK government and the Association of British Insurers announced a 5-year voluntary moratorium on the use of adverse genetic test results (HGC 2002: 122). In part this was an attempt to ensure that research programmes, particularly those related to cancer, were not adversely affected by the growing public concern following the scandals about the collection of tissues and bodily parts discussed earlier. The advantages to insurers of having access to genotypic as well as phenotypic information are clear. They decrease the costs to the insurance industry through a more efficient assessment of risks, and (the industry would argue) may also deter individuals from taking out expensive policies in the face of poor prognoses. There is also some concern that individuals with potential health problems would be impeded from seeking testing and put themselves (and perhaps their families) at risk (HGC 2002 Chapter 6). Also, the limits to current knowledge about the predictive power of genetic information were seen as not always well understood, particularly in popular perceptions of the possible implications for insurance (DTI 1998:3).

A major concern, particularly in the USA where health insurance is privatised and over 45 million do not hold any at all, centres on the creation of a genetic underclass (Andrews and Nelkin 2001). Applicants have already been denied medical or disability insurance for a range of inherited conditions including Huntingdon's Disease and type-one diabetes. Many of these will have no current symptoms, and it could be said that the judgements made had ignored an important aspect of genetic testing, namely that not all tests used in the USA (and currently under consideration in the UK) are the same. Some genetic tests are presymptomatic and diagnostic, as in the case of Huntingdon's Disease, and are relatively accurate. Others, for example for the breast cancer genes BRAC1 and BRAC2, are predictive and only signify a (albeit fairly strong) predisposition to breast cancer. Hence it is possible that carriers could be unfairly penalised if the disease did not develop, as may happen in up to 40% of cases (Pilnick 2002: 87). In any event, it is clear, as already noted, that many insurers and their clients fail to understand the significance of the genetic information they handle, and its implications for the future health of the insured. This particularly seems to be the case in differentiating between monogenic and polygenic disorders resulting in much genetic screening having speculative rather than diagnostic value (Bowring 2003: 208).

These difficulties also appear in the area of employment (HGAC 1999). Many employers see the advantage of accessing information about the future health of their employees as a way of ensuring continuity of employment. Whether or not an individual is asymptomatic when a genetic test is taken, and given that any result cannot predict ill health or age of onset with any certainty, an employer may still not wish to risk either sickness or premature retirement among employees. This would, it is argued, minimise the costs associated with absence through sickness, and the time wasted on training.

There is considerable public concern in the US about the use of genetic testing in employment (HGC 2002: 116 et seq) particularly in the context of recalcitrant employers not wishing to pay legitimate industrial injury claims or higher insurance premiums. The US Genetic Alliance, an umbrella organisation of people with genetic conditions, has collected case studies of discrimination among employees. It has highlighted the 'Terri Sergeant Case' concerning a woman who was fired after exhibiting early signs of a genetic condition because this would raise the costs of the company's group medical insurance. Thirteen percent of those with genetic conditions surveyed by the Genetic Alliance reported denial or dismissal from employment because of their conditions. Other disturbing cases include spouses of employees dying of genetic conditions offering children up for adoption in order that they may have access to medical insurance. About 17% of respondents had deliberately not revealed genetic information either to employers or insurers.

A further dimension concerns the concentration of certain genetic conditions among specific population groups such as the sickle-cell trait among Indians, African-Americans, Greeks and Turks. Despite the existence in the US of early diagnosis, and lack of any specific treatment, Du Pont began a decade long programme of screening all black workers and job applicants in 1972, initially without their knowledge or consent. Some organisations continue with this screening in the belief that this may make black workers more vulnerable to 'oxidising industrial chemicals, heavy physical exertion, or stress on the circulatory system' (Bowring 2003: 216). Others simply wish to ameliorate the effects on their insurance costs of paying the medical bills of offspring born to sickle-cell carriers. Under such circumstances, the line between employment discrimination and racism begins to blur.

DNA Profiling

A third area where the issue of consent has become increasingly problematic saw its beginnings in 1983 when a technique originally developed to identify the markers associated with familial disorders was adapted to help track down a rapist. It is now possible for forensic scientists to create a DNA 'fingerprint' from a saliva sample, a spot of blood, a hair root, or a semen stain. This methodology became commonplace in a number of important areas outside that of criminal forensics including the identification of the remains of partially dismembered dead bodies in war situations, the hunt for missing persons after natural disasters, or the resolution of disputed parentage in contentious custody cases. It has been described as a major contributor to 'surveillance creep', particularly following the policy instituted by the UK government in 1989 to use the DNA fingerprint test on immigrants seeking to prove that they have relatives already living in Britain. This practice spread to Canada, the US and elsewhere in the early 1990s (Nelkin and Andrews 1999: 200 et seq).

However, it is in the field of crime prevention and detection that the possibilities of surveillance creep and the threat to the concept of informed consent are most clearly delineated. As DNA techniques improve and their cost decline, there have been calls for the establishment of a population wide profile database in the interests of more efficient policing (Gammon cited in Nelkin and Andrews 1999: 202). In the US a national programme to assist federal, state and local law enforcement agencies was established in 1993 linking the DNA profiles of convicts gathered by scattered law enforcement DNA labs, and called CODIS (Combined DNA Index Systems). The UK National DNA database contains 1.5 million DNA profiles. The majority of samples have been collected from individuals who have been convicted or suspected of involvement in a crime, or from volunteers who have given samples for elimination purposes (HGC 2002, Chapter 9).

These forensic databases store numerical representations of 10 regions of rDNA lying in the 'junk' DNA inherited from both parents. They are therefore profiles that are unique to an individual except in the case of identical twins. These profiles differ from the tissue samples from which they are derived in that they normally contain no predictive information (for example about the future health of an individual), and can only generate information about parentage when compared with other identifiable profiles. The samples from which they are drawn, on the other hand, can be used to generate personal genetic information, but do not form part of the National DNA database. Under the UK Criminal Justice and Police Act 2001, samples collected as the result of a police investigation can now be retained indefinitely, even if the suspect has been acquitted of the crime for which they were first taken. In addition the law allows the police to take bodily samples without consent from those suspected of a variety of offences, including many less serious than murder or sexual offences. The addition of profiles derived from such samples to the National DNA database, unlike other aspects of data collection, is not covered by the Data Protection Act 1998, and is irrevocable.

There appears to be an increasing divergence between public opinion and police practice in this area of civil liberties and the relationship between the citizen and the state (HGC 2002: 149). It is not helped by issues surrounding the validity and reliability of the collection, analysis and use of DNA evidence (Lander 1994, Lynch and Jasanoff 1998). Too often cases brought to court have been found deficient in one or another of these areas. It is also compounded by concerns about privacy and confidentiality. For example there are, according to Nelkin and Andrews (1999: 198), over 19,000 law enforcement agencies and over 51,000 criminal justice agencies worldwide with direct access to the Crime Information Centre maintained by the FBI. The possibility of ensnaring the innocent through the rapid expansion of DNA fingerprinting over the last two decades seems likely to increase, as does the degree of surveillance of the innocent activities of the wider population.

Bioinformatics

The advent in the early 1980s of DNA sequencing as part of the Human Genome Project (see Chapter 2 above), resulted in an exponential growth of molecular sequence data accumulated in various databases around the world. These included Genbank, EMBL (European Molecular Biology Laboratory nucleotide sequence database), DDJB (DNA Databank of Japan), PIR (Protein Information Resource) and SWISS-PROT. They were accompanied by increasingly more complex computational methods for data retrieval and analysis, aimed at establishing sequence similarities, and structural and functional predictions (Courteau 1991). These databases also had the purpose to pool data from disparate sources so information could be readily shared in the production of the maps that were the goal of HGP. As Cook-Degan (1994: 292) notes, a new field of bioinformatics developed, linking biology inextricably with computers and mathematical techniques, to construct theories of biological structure and function.

However, this development was not without its problems. Databases from around the world were often not compatible as they were differentially structured, used incompatible hardware and software, and contained different data (Fields 1992). They were also located within different organisational and management regimes. One clear example of the difficulties faced was revealed when YAC (Yeast Artificial Chromosome) data transmitted from French to US laboratories revealed wide variations in reliability and even usefulness. In particular, a number of 'chimera' had been included as part of a mega-YAC library produced by CEPH/ Genethon in Paris (Anderson 1993; Evans 1993).

Further difficulties were posed by the leakage of funding from biology to bioinformatics as these technical problems became apparent and needed to be resolved. As Cantor (1992: 106) notes funding such activities had played only a small part in traditional biological research, and had been difficult to come by. But the HGP should have been spending at least one half its multi-million dollar budget on the development of these new techniques and technologies.

In parallel, a new breed of scientist began to evolve shifting 'power into the hands of those with mathematical aptitudes and computer savvy' (Cook-Degan 1994: 293–294). Cantor (1992: 109) predicted the heavy dependence by biologists on the computer if the HGP was to be completed on time. In particular, he saw the disappearance of the laboratory notebook so beloved of his British colleague John Sulston (2002). One way forward was the electronic community system or collaboratory developed by scientists studying a model organism for the HGP, the nematode worm *C. elegans* (Schatz 1991). Lederberg and Uncapher (1989): described a collaboratory as using the new information and computing technologies in a

> combination of technology, tools and infrastructure that allows scientists to work with remote facilities (co-laboratory) and each other (collaborat-ory) as if they were co-located and effectively interfaced… (p. 3)

'The collaboratory will provide a seamless access to colleagues, instruments, data, information and knowledge' (p. 6).

This approach ignores a key element in any attempt to replicate data for local use, namely the role of tacit or craft knowledge. Typically any form of replication in science requires the communication of this kind of knowledge if it is to be successfully accomplished (Collins 1985, Latour and Woolgar 1979), and this seems particularly to be the case in biological research (McCain 1991). Simply accessing a datum from a database generated elsewhere may not be sufficient to allow immediate use at the local level. As Hine (1998: 42) suggests, the databases should be seen as an integral part of the characterisation of what constitutes a gene since the 'observer, the machine and the object of study are mutually constructed in laboratory discourse'. Whatever the limitations towards the end of the last century, the present ubiquity of the Internet has according to Kanehisa and Bork (2003: 305) 'transformed databases, access to data, publications and other aspects of information infrastructure' in the post sequence era.

However, some difficulties still remain as primary databases of sequence data continue to expand as a result of the introduction of high throughput sequencing methods. The results are not always up to date or well annotated, mainly as a consequence of now being produced with a throughput of about 6,000 times that of early prototypes (Hood and Galas 2003: 445). This has led to an increasing reliance on secondary databases that contain signatures of protein domains and functional sites, facilitating an understanding of functions and utilities at the molecular, cellular level of organisms rather than individual genes or proteins (Kanehisa and Bork 2003: 306). The further developments of these form the precursors to understanding the basic principles of the higher complexity of biological systems using bioinformatics. 'The ultimate integration of biological databases will be a computer representation of living cells and organisms, whereby any aspect of biology can be examined computationally (Kanehisa and Bork 2003: 309).' This information-based view of biology may only come about after a significant change in the culture of science, which on past evidence may be more difficult to achieve (Glasner 1996).

One reason lies in the process of standardising bioinformatic tools and data. As Fujimura (1999) notes, standardisation has two distinct meanings. Firstly standardised data allows laboratories to compare and contrast outcomes produced in different sites. Secondly after having been accepted as standardised, the data begin to shape future outcomes and commitments. They are, therefore, not in any sense any longer neutral, raising questions about situatedness, historical contingency and authority (Haraway 1991). Standardised data become boundary objects (Star and Griesemer 1989: 393) that are both plastic enough to adapt to local needs and constraints of the several parties employing them, yet robust enough to maintain a common identity across sites. As Bowker and Star (1999: 14) point out, citing such examples as QWERTY, Lotus 123, DOS and VHS, there is also no natural law that the 'best' standard shall win. Increasingly, therefore, the contextual nature of stored genomic data is being recognised, as functional genomics becomes the dominant paradigm and the goal of research becomes more than simply cataloguing all the genes, and the information about their functions.

Conclusion

In this chapter we have explored the range of concerns associated with the management of genetic information resulting from the techno-scientific advances in functional genomics in the generation and storage of personal and other data. In particular we have discussed the development of biobanks in the UK and abroad. This has raised issues surrounding difficulties in obtaining informed consent for the use of data, and the impact on individuals and families. Policy issues about the relationships between public and commercial ownership have also been investigated especially when data privacy, and the use of samples and data for unrelated research or the development of therapeutic downstream applications, has proved potentially contentious. These have been exacerbated by worries voiced by the public, particularly in the UK, concerning the collection and storage of body parts and tissues by various hospitals in the recent past.

We have also discussed how issues of privacy and anonymity impact outside conventional health related areas of concern. The possibilities of discrimination in insurance and employment were explored, particularly given the relatively poor understanding that many practitioners have of the meaning of the data on which their decisions are made. We also briefly touched on the use of personal genetic information in forensics and police work, where the contextual and situated nature of knowledge was raised. This was developed further in our analysis of the growing importance of large-scale genetic databases for pushing forward the frontiers in the post genomic age. Traditional laboratory based biological techniques are being rapidly replaced by the introduction of multidisciplinary teams using high throughput techniques and multiple database analysis to create a more systems-based approach to understanding functional genomics. However, these advances raise issues for the day-to-day organisation of the community of science, and for the processes by which techniques, and the data they produce, are standardised across the community.

The issues and concerns surrounding the management of genetic information have global as well as local implications. We have attempted to clarify these by opening the 'black box' called personal genetic data using insights from a number of mainly qualitative studies, and some of their results also form parts of the remaining chapters in this book. In the next chapter we approach another 'black box' and discuss the more contentious (at the time of writing) issue of genetically modified crops and foods.

Chapter 5

‘Frankenstein’ Foods, or the Revenge of the Genetically Modified Potatoes

The latter years of the twentieth century have witnessed a change in focus regarding the possible effects of what we eat on human health and the environment. So long as modern farming methods utilised ‘craft’ technologies based on selective breeding to produce our perfect red tomatoes, or our flawless green and shiny apples, the public seemed happy to assume that any risks or hazards were controllable and unlikely to be catastrophic. However, with the advent of genetic modification, and its introduction into food production, public opinion, especially in Europe, appears to have radically altered. It now seems that the focus of concern is as much on the safety of the new genetic techniques as it is on the engineered crops they have been used to produce.

In this chapter, we will discuss these important cultural shifts, and use the resources offered by science and technology studies to see to what extent we can illuminate what, for many, is very much a technological ‘black box’. In doing so, we locate the discussion in the global experience of late modernity where GM crops and foods are rapidly rising on the agenda of the G8 nations. We follow this by revisiting the recent history of the controversy over genetically modified food and agriculture, and then analyse the role of scientific experts, and how their views relate to those of the lay public. We follow by exploring the role of the State in drawing the policy boundaries within which it unfolded. In conclusion, we explore the involvement of the ordinary citizen in decision-making.

Barbarism Modernised: the Age of Risk

Natural hazards have always made society a dangerous place whether in ancient times, or later, in predominantly agricultural economies (Giddens 1998: 27). These were also widespread, though their causes were often attributed to God or Fate. However, following industrialisation in the nineteenth century, and the rise of science and technology, society in the mid-twentieth century has become characterised by the ubiquity of its crises: apocalypse has now become banal (Giddens 1991: 183–184). According to Ulrich Beck (1992), a major reason for this perception comes from the fact that, while in previous epochs science and technology were part of society’s attempt at mastery over nature so as to control hazards, experiments were generally confined to the laboratory bench. Now, society itself has become part of a global laboratory, and the risks have replaced hazards as the global threat. Whereas, initially those responsible for creating hazards could be identified, especially since the unquestioning authority of science was embedded in the wider culture, today's risks escape direct perception as they appear in toxins in foodstuffs, or in fallout from nuclear accidents, and there is no-one to take responsibility. Their victims often do not detect them until it is too late, and some of their consequences only become visible in future generations (Beck 1998).

The result is that the claims of science to a monopoly of expertise have become more difficult to sustain in the face of competing claims, interests and viewpoints from the 'agents of modernity' (the producers of risk) and the injured parties. An obvious, recent example is the link between new variant Creutzfeld-Jakob Disease and so-called 'Mad Cow Disease' or BSE. Here we do not see a simple scientific problem resolvable at the laboratory bench. Instead, as Wynne (1996a) observes, the issues are co-constructed by British relationships with the European Union and its regulations, the apparent proximity of government to private industrial interests, fast-eroding public confidence in official bodies, and a 'distinct whiff' of political control of science, alongside painstaking medical and scientific research. Society also faces the uncertainty of whether or not the thousands of innocent victims of 'beefgate' (*The Independent* quoted in Beck 1998:10), who enjoyed a diet of hamburgers in recent years, are already infected with a life-threatening disease for which as yet there is no cure. They are part of a living laboratory, waiting to see over the next generation if this unwanted experiment will end in disaster (Cousens et al 1997: 197). While these threats are real, the perception of risks is as much to do with a breakdown of trust resulting from an increased dependency in modern society on expert-led institutions, as it is to any direct link between CJD and BSE.

In the Introduction to the English edition of Beck's *Risk Society* (1992: 1–8), Lash and Wynne relate his discussion to recent advances in the sociology of science and technology, and in particular to the results of studies undertaken under the Economic and Social Research Council initiative on the Public Understanding of Science (Irwin and Wynne 1996). They suggest that society operates with an idealised model of the risk system, which gives undue weight to laboratory knowledge without recognising the contextually narrow nature of its production, and then proceeds to give it the force of prescription. The example they use is of the reaction by the British government to the alleged health hazards associated with herbicides by farm workers. The Government turned to the Pesticides Advisory Committee (PAC), which reviewed the scientific literature on the laboratory toxicology of the chemicals in question, and concluded that no risk existed. An even fatter dossier of cases of medical harm was dismissed as 'merely anecdotal, uncontrolled non-knowledge'. When pushed further, with even more examples, the PAC admitted that there was no risk as long as the herbicide was produced and used under the correct (laboratory) conditions. However, they failed to recognise what all farm workers knew: that instructions for use were frequently obliterated or lost, the proper spraying equipment was often unavailable, protective clothing was often inadequate, and weather conditions were frequently ignored in the pressure to get the spraying done.

Experts are given the role of defining agendas in the risk arena, but as has been seen time and again, and increasingly, seem unaware of what sociologists and others have discovered by their research. Lash and Wynne summarise this in three points. Firstly, risks, however real, are socially located; secondly their magnitude is a direct function of social as well as physical processes; and thirdly, the primary risks are often 'alien, obscure and inaccessible' to those most likely to experience them. The relationship between scientific expertise, generated through laboratory knowledge, and the contextually generated lay understanding of the application of knowledge, is perhaps nowhere more chillingly illustrated than in the case of supposed Gulf War Syndrome. Here, organophosphate pesticides, similar to sheep dips and chemically related to nerve gas, were widely used by British troops in the Gulf. They were apparently kept in ignorance of its potential consequences, as indeed was Parliament, afterwards, by the Ministry of Defence (Fairhall 1997: 6). Malathion, used in very small quantities to kill hair infestations in children, was applied in large quantities and without the benefit of protective clothing, to delouse prisoners of war. The resulting symptoms of this pesticide misuse bear close similarities to those experienced by the farm workers using agricultural organophosphates described by Lash and Wynne. The effects have already resulted in disability and death, and some suggestion of transmission across generations to the children of war veterans. The case of genetically modified (GM) food has not developed against a backdrop of severe illness and loss of life in the way that the controversies over CJD-BSE and Gulf War Syndrome have. However, the potentially hazardous nature of introducing genetically modified organisms (GMOs) into the world has begun to create a debate with similar sociological characteristics. It may, therefore, be helpful to begin with a brief outline of the key events as they occurred in Britain (but see also Adam (2000), Levidow (1999) and a special issue of *Sociological Research Online* (SRO 1999)).

The Genetically Modified Food Controversy

The study of scientific controversies has a long history in the social studies of science and technology, and a good overview is provided by Nelkin (1994; see also Nelkin 1982), while a more general account is to be found in Englehart and Caplan (1987), and the distinctively British tradition of the 'Bath School' approach can be found in Collins (1981). Nelkin (1994) suggests that controversies arise in a variety of contexts, for example, when questions of social equity clash with those of economic efficiency as in the siting of nuclear waste facilities – the Not In My Back Yard (NIMBY) syndrome. They may also stem from fears of risks associated with contamination such as resulted from the disaster at Chernobyl. However, they can equally arise from the apparent curtailment of individual freedoms as in the case of the hazards of smoking to health, or the violation of traditional values as with the development of *in vitro* fertilisation techniques.

Following a long history of directed plant breeding, deriving from Mendel's experiments on peas in the nineteenth century, the first genetically modified plants – tobaccos – were developed as early as 1983, and the first foods – cereals – in 1990. The use of GMOs for the development of herbicide-resistant crop strains such as soya, and their introduction into the food chain, began to arouse controversy in the 1990s in the context of massive public concern after the identification of a possible CJD-BSE link, and alongside a series of other food-related safety concerns surrounding outbreaks of (sometimes lethal) *E.coli*/0157. Together with the introduction by a major supermarket of genetically modified tomato paste, which was on sale and labelled as such for eighteen months without causing undue concern, came calls from consumer organisations, and non-governmental organisations (NGOs) such as Friends of the Earth and Greenpeace, for better labelling of foods to improve and facilitate consumer choice (see *inter alia* Hamilton 1998). Organisations such as schools and supermarkets even banned the use of foods identified as containing GM ingredients some six months after concerns had been expressed by NGOs, when it became clear that legislation both in Britain and Europe concerning labelling was inadequate to deal with *processed* foods such as biscuits or pies, where the GM input had occurred at the very start of the food chain through, for example, the mixing of GM and non-GM products such as soya, prior to their processing and export from the primary producer, the USA.

This complex set of inter-related difficulties was further compounded by a report from the prestigious Rowett Research Institute in October 1998. Dr Arpad Pusztai, a senior member of its staff, appeared to have shown that potatoes, genetically modified to contain a snowdrop gene, adversely affected the organs and metabolism of rats to which they were fed. While his unpublished results did no more than raise questions of safety for future research, their announcement, made in a television documentary, caused a political storm (see, for example, Driscoll and Carr-Brown, 1999), and Dr Pusztai was dismissed following an internal audit. His research became part of the media debate over the dangers to the public of eating any GM food, and finally the Royal Society produced a Report on his experiments, peer-reviewed by half a dozen eminent statesman of science, which concluded that no reliable or convincing evidence for his claims had been provided (Royal Society 1999).

The controversy then widened from a clear public concern about foods, to reservations about the possible detrimental effects of growing GM herbicide resistant crops in proximity to those not so modified. In particular, government-approved trials in several farms across Britain were found to be only several hundred yards from neighbouring farms producing officially recognised organically grown crops. The danger of cross-pollination meant that these farmers were in danger of losing their recognition, and having their livelihoods seriously jeopardised. A US study by researchers at Cornell University, reported by Kleiner (1999) in *The New Scientist*, suggested that in their experiment, one half of monarch butterfly caterpillars fed on leaves of milkweed (their only food), dusted with pollen from GM corn, died within four days. However, their research had also not been peer-reviewed, and was instantly dismissed by experts from industry and the US Environmental Protection Agency. In Britain, English Nature vociferously expressed its concern about the possible ill effects of any large-scale trials of GM herbicide resistant crops on adjacent flora which provide the habitat for a rich diversity of birds and insects. These concerns were dismissed by the Nuffield Council on Bioethics (1999), while the need for caution was expressed in an Interim Statement issued by the BMA (1999). The questions raised by the various parties were summarised in a most public fashion by the Prince of Wales in an article in *The Daily Mail* on 1 June 1999, generating even wider debate about the respective roles and responsibilities of science, government and industry.

In 2001, the UK government established an Agricultural and Environment Biotechnology Commission (AEBC) to advise it on biotechnology issues.

Following its first report, *Crops on Trial* (AEBC 2001) the government proposed a moratorium on the commercial growing of GM crops in the UK until the completion of a series of farm scale evaluations (FSEs) of the impact of genetically modified herbicide tolerant oilseed rape, sugar and fodder beet, and forage maize on farmland biodiversity. In April 2002, the AEBC suggested that a country wide public debate be instigated on the commercialisation of GM crops prior to the publication of the results of the FSEs in the summer of 2003. This is discussed later in the chapter.

As this controversy unfolded in the UK, the world's largest supplier of GM food and crops, the United States of America, has continued to press for their acceptance around the world. However, as it became clear that a ready market, particularly in Europe, and increasingly in some less developed nations, is not guaranteed, the initial acceptance by US farmers of GM products became increasingly to be questioned. One reason appears to be that the products were oversold as being more likely to increase yields significantly and hence increase profits. Recent reports suggest that the contrary may be the case with yields for Monsanto's Roundup Ready GM crops either yet to show an increase, in the case of maize, or to actually be reduced, in the cases of soya and oilseed rape (Soil Association 2002, Chapter 3). Another is the problem of cross contamination mentioned earlier, with GM pollen spreading to non-GM areas (including Organic Farms) via the wind, insects, floods and farm machinery (Steinbrecher 2001). A third reason is that of the development of herbicide tolerance, particularly in oilseed rape. Because of its physical characteristics, it is particularly susceptible to dropping seeds during harvesting which then germinate anything up to ten years later. The resultant 'volunteers' embedded in future crops grown on the land and are often resistant to the herbicides these crops require, increasing the costs to the farmer (Soil Association 2002, Chapter 6). Some even become resistant to several varieties of herbicides, and have been labelled 'superweeds'. The US is however by no means as embroiled in a controversy over GM crops and food as a result of these growing issues compared with Britain and Europe.

Controversies, according to Mazur (1981), typically developed in three stages: a warning stage, a public stage, and a mass movement. In the case of the introduction of GM crops and food in the UK, the warning stage is best illustrated by the actions of a whistle blower (Arpad Pusztai), and an interest group (English Nature). The problems highlighted by a growing minority of farmers in the US also correspond to this stage. The mass clearance of GM-based foods from supermarket shelves, and the issuing of reports by the BMA and the Nuffield Council on Bioethics, constitutes a good example of the second, public, stage, while the extent to which the British public entertain doubts about the safety of GM foods in the long term will determine whether a mass movement with lasting effects is created. The study of controversies facilitates insight into the policy-making process, helps promote a greater understanding of the roles of experts, illuminates ways in which non-scientists can become involved in decision-making, and clarifies the social and political nature of the scientific enterprise.

Science Policy and the Regulation of Risk

A key element of the policy-making process is the development of appropriate laws and regulatory agencies to ensure that new scientific and technological advances do not pose unnecessary risks to the public. The adoption in the late 1970s, by the statutory Genetic Manipulation Advisory Group (GMAG), of a hazard-tree rather than phyllogenic-based scheme for reconceptualising the risks associated with the regulation of genetic manipulation techniques in Britain (Bennett, Glasner and Travis 1986) followed a world-wide moratorium in 1974 arising from public concern in the USA and Europe (Krimsky 1982: Wright 1994). Initially, scientists were only able to identify degrees of risk in terms of the closeness of any manipulated genetic material to humankind. Hence manipulating the genes in a plant was likely to be less hazardous than, for example, using a mammal. However, the net result was seen by scientists as too inhibiting of research, and following the relaxation of the moratorium in the USA, it became necessary to find an alternative method of calculating the hypothetically possible risks concerned. The solution was to adopt a hazard-tree approach which allowed calculation of risks to be made at each stage of a project. This opened the way to relating the riskiness of any experiment to a suitable laboratory environment, and allowed even the most potentially hazardous of research projects to take place, for example, in facilities such as Porton Down. In this way, the assessment of uncertainty was transformed into merely following a bureaucratic routine, with the eventual abolition of GMAG as a regulatory body, and the incorporation of many of its functions into the Health and Safety Executive. This routinisation of the risk analysis process is seen by Jasanoff (1995) as part of the repertoire of rhetorical devices used by science advisers to ensure that political decisions are not made which could be detrimental to their desire for unfettered pursuit of knowledge.

Controversies can, however, also be seen as sites of 'social learning' through an informal process of technological risk assessment (Mazur 1981, Rip 1986), in contrast to the formalised processes developed by official regulatory agencies. Cambrosio and Limoges (1991) go further to suggest that controversies are indeed the *central* element of any social assessment of technological risk. Wynne (1995a), using the example of the ways in which the European Commission facilitated the development of the European biotechnology industry, argues that the failure of draft legislation on the regulation of animal growth hormones to get Commission approval in 1985 (in spite of significant backing from the scientific establishment) stemmed from a lack of recognition that the required socio-technical conditions for its implementation were never likely to be realised in all member states. The resulting controversy stemmed not from a debate between science and anti-science, but from two different, but equally legitimate, views on how a risk assessment culture operated in the real world. For the scientists, all that was required was a set of formal procedures to be carried out in a prearranged manner. For the interest groups, whose influence scuppered the legislation, it was clear that the social policing necessary to ensure that the regulations were correctly applied at all times, by farmers in all the member states from Greece to Italy, was never likely to be put in place. The development of a controversy over animal growth hormones in cattle has, therefore, had the effect of illuminating a significant area of real concern in Europe.

These resources can be used in the context of the continuing debate about genetically modified food and crops. For example, there has been growing public concern about the lack of information available about contents on food packaging. The Food Advisory Committee in 1993 established guidelines for labelling which were accepted by the British Medical Association and the National Farmers Union on the grounds that the precautionary principle must be the right way to proceed. Supermarkets were dismayed mainly because of the likely cost of which not all would necessarily pass to the customer. Consumer organisations and other interest groups felt that they did not go far enough, and without the force of law, were unlikely to be consistently applied. So, the controversy began. The decision to pass legislation requiring labelling of food ingredients was confirmed in 1995/6, and in the Spring of 1999, a new and more thoroughgoing law was announced which recognised, in particular, the need to identify any ingredients (over 1%) which had been genetically modified. While the new proposals still did not go far enough to satisfy some radical environmental groups, they were seen as politically and socially acceptable since they appeared to act in the best interests of the consumer by facilitating choice.

However, it was slowly becoming apparent that, since a major ingredient of supermarket food in Britain was genetically engineered soya and maize grown primarily in the USA, the apparently straightforward matter of identifying whether it was genetically modified was to be problematic. The difficulty revolved around the fact that, working under a more liberal regulatory regime, American farmers had for almost a decade mixed the genetically modified beans or maize with those grown without the benefit of scientific intervention, prior to processing them for export around the world. Soya is an ingredient in about two thirds of the processed foods, such as ready-to-eat meals, consumed in Britain. It is also a constituent of cooking oil, and with starch from maize, the emulsifier, lecithin. As a result, the requirement to label a food as including genetically modified ingredients has proved impossible to fulfil; the processing undergone prior to export results in a new product which provides no recognisable evidence of the modification of its constituent parts. In addition, the proposed legislation completely ignored another fatal complication, since even foodstuffs sourced from non-USA outlets could include a proportion of US grown, and therefore probably modified, ingredients.

Hence this case study illustrates how regulatory agencies and laws are developed in order to help establish and control risks enabling the public to make informed choices. But new regulatory regimes involving labelling appear to have benefited from the application by industrial science advisers of just those rhetorical devices identified by Jasanoff (1995) as facilitating rather than inhibiting further research on GM food ingredients since the modifications take place prior to processing. The labelling debate has also helped identify the major protagonists, industry, government, the NGOs, and shown the importance of social and political concerns in the decision on what, on the surface, appears simply to be scientific advice on regulatory policy. Science alone, it seems, is unable to establish the *definitive* basis for improving consumer choice. The case study also shows how, in spite of expert scientific advice, the framing of new laws and regulations in a 'risk society' cannot guarantee that the unimaginable will not arise. In this way, it clarifies the importance of controversies in facilitating the process of social learning in technological risk assessment.

The Tragedy of Expertise

Scientists have throughout history used their status as 'experts' to both draw a line between their own work and that produced by quacks and charlatans (Collins and Pinch 1982), and to defend themselves against the consequences of uncritical consumption of their research by non-scientists such as people in business and government – the 'lay public' (Gieryn 1983). In a period of 'reflexive modernisation' (Beck, Giddens and Lash 1994), the institutions of risk society are opened to criticism as never before, and scientists' own pronouncements about the nature of the dangers science itself creates are increasingly revealed as lacking legitimation. Sociologists have noted for some time that disagreements between experts provide useful insights into how scientific knowledge is constructed and evaluated (Irwin 1995). Barnes and Edge (1982: 237) in their review of the early literature, suggest that the credibility of a scientific expertise rests as much with the contexts in which it is offered, as it does with the rationality of the arguments. They describe the recognition of this ubiquitous contingency as the 'tragedy of expertise'.

In their study of medical and scientific professionals in the new genetics, Kerr, Cunningham-Burley and Amos (1997) suggest that this tragedy is still unfolding. Their respondents held the view that rational scientific knowledge was a 'gold standard' clearly demarcating good, and value free, research from illogical, or politically distorted, opinion, which they paternalistically attributed to an undifferentiated lay public. However, these experts went further in appearing to promote objective advice about the risks of the new genetics, while simultaneously disguising the extent of their own social location and vested interests. This has recently become part of the GM controversy as well with the publication in the national press of the links between membership of government advisory committees, including ACRE and the Advisory Committee on Novel Foods and Process (NCNFP), and major biotechnology companies such as Monsanto, Zeneca and Novartis. In the case of ACRE and NCNFP, it was suggested that one half the members have such links (Nuki 1999). These were not made clear to the public when the committees initially proffered their advice to government on the regulation of GM foods and crops.

Gilbert and Mulkay (1984) identified a difference, based upon public and private pronouncements by scientists, which helps to explain how experts' contrasting linguistic repertoires are themselves socially located. They suggested that two distinct repertoires exist, and scientists are clear when one or the other is to be used. The first recognises the contextual nature of knowledge production, is used informally, for example in conversation between scientists in the laboratory, and is called the 'contingent repertoire'. However, when communicating formally, as in published reports and papers in academic journals, or communicating with the public, scientists invariably resort to using an 'empiricist repertoire'. Here events, actions, and beliefs become a neutralised medium through which the empirical truth emerges, and science is reconstructed as a rational and value-free activity. This implicit presupposition of the politically and morally neutral value of science contributes to constructing a particular identity of 'science-in-general' (Michael 1996: 105) which then becomes a vital resource in enrolling the public into accepting the validity of scientific expertise. It can also be argued that those most prominent experts are scientists who are also furthest from day-to-day research at the laboratory bench (Glasner and Rothman 1999). Hence their pronouncements are made with a degree of confidence which those much closer to its contingent nature might find more difficult to sustain; distance may effectively legitimate certainty (Collins 1988: 726).

One reason for studying controversies is that they force protagonists to clarify the often tacit and un-articulated assumptions that underpin their arguments. The Pusztai case mentioned earlier, and widely publicised in the media (see, for example, Radford 1999), throws light on the significance of using appropriate repertoires to legitimate expertise, and clearly highlights for the wider public the tacit assumptions underlying what constitutes certified scientific knowledge. Dr Pusztai stood accused by many leading experts of having made his research findings public without first going through the normal channels and publishing them in a peer-reviewed specialist journal. As a result, it was argued, the validity of his findings could not be properly evaluated, and, given their controversial nature, the publicity had thrown his employers, the prestigious Rowett Research Institute, into disrepute. The issue was further clouded by the need for Pusztai to seek further funding for his work, suggesting a mix of motives similar to those found in a previous case, that of Cold Fusion research, where the results of a potentially ground-breaking experiment were also made public at a televised press conference (Close 1992). It is ironic that, as mentioned in Chapter 3, no such concerns were voiced when President Clinton and Prime Minister Blair publicly announced the completion of the project to map the human genome even though it was unpublished and incomplete. This violation of scientific protocol has the effect both of making more complex the accepted linear method of scientific communication, and of adding to rather than simplifying the confusion about how valid the science really was (Lewenstein 1995).

Dr Pusztai was suspended, pending an internal audit of his work. Following its failure to confirm his findings, the Institute allowed Dr Pusztai to publish an alternative report on the internet. During this time he was given very public support by a group of twenty international scientists, who suggested that it was too early to wholly dismiss his claims, and that further research might well establish the validity of his conclusions. When his report became publicly available, one national newspaper requested a 'peer review' from a respected expert toxicologist, and subsequently printed his critical findings (Sanders, in Conner 1999). The matter came to a head with Dr Pusztai's data being submitted to a distinguished panel of experts from The Royal Society, who concluded that the results of any further tests for GM food safety should not be published until they had been appropriately peer reviewed (Royal Society 1999). *The Lancet*, while issuing an editorial disclaimer, felt it appropriate to publish at least some of his findings later in the year. The reaction from the leaders of the scientific community remained dismissive. This effectively closed the debate and re-established, at least in the short term, the hegemony of the scientific establishment over the certification of scientific knowledge (Collins 1985: 142 et seq).

These events also vividly illustrate the important differences between linguistic repertoires in the legitimation process, showing how both the initial televisual 'publication' of the results, and the subsequent media-inspired 'peer review' process, were couched in a discourse characterised by the contingent repertoire identified by Gilbert and Mulkay. This had the effect of denying scientific validity to either side, and served only to further fuel the flames, and muddy the waters, of the controversy about the potential risks of GM foods. Closure to the debate only became possible through the intervention of The Royal Society, the British scientific institution with the highest status. Its Report, based on the use of orthodox review procedures, and couched in careful and unambiguous language, thereby using the empiricist repertoire which distanced the authors from its conclusions, rejected Dr Pusztai's data as inadequate. However, it was also an example of how actor-networks (Callon 1986; Latour 1987) function to enrol and mobilise expert opinion and identify as dependent a lay public in mending ruptures to its legitimating fabric. For the scientific establishment, at least, the 'tragedy' had been averted through this 'simplifying' intervention into an increasingly complex controversy involving television, the press, the scientific media including official journals, and the world-wide web.

Involving the Public

The need for greater citizen involvement in policy decision-making is not solely limited to developments in science and technology (Pateman 1970, Young 1990, Cochrane 1996) although the results of studies under the overall rubric of 'the public understanding of science' (PUS) have brought them, according to Irwin (1995), into the foreground for STS (see, for example, Eden 1996). Dorothy Nelkin (1975) noted that to define decisions as technical, as normally occurs with each new advance in science and technology, rather than political, effectively dis-enfranchises the public at large. Recent studies of PUS (for example, Irwin and Wynne 1996), have established that scientists utilise what Wynne (1991, 1995) describes as a 'deficit' model of scientific knowledge when evaluating lay contributions to technical debates. This assumes that public reservations about science and technology are predicated upon ignorance, so that better science education will necessarily lead the public to a greater understanding and more objective appreciation of its contribution to social progress. However, by default, it also implies that the only legitimate carriers of such knowledge are those already trained as scientists themselves, and that what constitutes knowledge is itself universal and uncontentious (Michael 1996).

The 'deficit model' of scientific knowledge highlights the recognition that 'framing' what constitutes legitimate science is a socially constructed process involving numerous groups and institutions in society. Until the recent past, responsibility for reproducing the hegemonic view has rested with a scientific establishment enjoying the active support of society at large (Beck 1992). In the sociology of science, this has been represented by the systems approach of Robert

Merton, as discussed in Chapters 2 and 3, who famously regarded the scientific norms of communalism, universalism, disinterestedness, and organised scepticism, (known as the CUDOS syndrome) as a vital ingredient of the democratic process (Merton 1973). The constructivist 'turn', building on the sociology of knowledge, social interactionism, and ethno-methodology in the 1970s (Mulkay 1979), paved the way for STS to significantly contribute to the wider theoretical debates in sociology as a whole (see, for example, Velody and Williams 1998).

The opportunities for acceptable participation in the policy-making processes of science and technology are now seen as severely circumscribed by the institutions within which they occur. Much work has been done on alternative ways to involve the public in overcoming this democratic deficit. Early attempts, particularly based on survey research, suffered from confusing the citizen with the consumer, and confounding the discourse of market competition with that of rights and responsibilities (Irwin 1995). More imaginative ways of avoiding this confusion have included focus groups, public meetings and conferences, citizens' panels and juries, and deliberative polls (Seargeant and Steele 1998), although each have their strengths and weaknesses (Coote and Lenaghan 1997). These are discussed further in Chapter 8.

In November 1994, the Science Museum organised a Consensus Conference on Plant Biotechnology, the first of its kind in Britain (Joss and Durant 1995, Barns 1996, Fixdal 1997, Purdue 1999). A lay panel was constituted to question witnesses from science, industry and NGOs, in order to arrive at a view about the future of this contentious technology. However, according to one observer, the Conference suffered from a deep ambivalence about whether it was there to facilitate the decision-making process, or to simply be part of an ongoing consultation exercise (Purdue 1999: 96). As a result the expert-lay divide became further entrenched, and, what on the surface was delivered as a consensual outcome, was actually unconnected to the policy process and therefore of little practical value.

The Citizen Foresight Project was an attempt by the London Centre for Governance Innovation and Science and the Genetics Forum (LCGIS 1998) to involve the public while addressing some of the issues raised in the consensus conference. It focused on the future of agriculture and the food system, including questions on sustainable agriculture, organic food, genetically modified organisms and public confidence. Unlike in the jury model, the twelve randomly selected 'panelists' (as they were deliberately called) from the Brighton area themselves set out the issues they wanted to explore. Witnesses were assessed by a Stakeholder Panel of key interest groups to ensure a balanced view, and the conclusions were drawn up by the panelists themselves. The Report was launched at the Parliamentary Environment Group at the House of Commons in an attempt to directly influence policy-makers. In this way the organisers hoped to both contribute to offsetting the democratic deficit, albeit for only a small number, ensure that expert credibility was co-constructed rather than imposed, and forefront the importance of public involvement in framing the questions as well as suggesting the way forward.

The debate about the possible commercialisation of GM crops proposed by AEBC in April 2002 was also predicated on the premise that the government would prefer its decision-making to be informed by a thorough public debate involving all the key stakeholders. This would 'clarify and advance public views' in the context of the last few years of strong publicity in the media (AEBC 2002: 2). The core programmes was to involve people from the grass roots in local communities based on meetings, focus groups and films. In addition, a number of additional elements could be added including one or more consensus conferences, or an interactive television debate. The funding from the government was not sufficient to provide for more than a small proportion of these recommendations and then only after an angry exchange of letters between the chairman of the AEBC and the Secretary of State for Environment, Food and Rural Affairs, Margaret Beckett. The outcome was a six-week round of debates which began in June 2003 at the National Exhibition Centre in Birmingham. According to newspaper reports, the events were to be seen as 'a unique experiment to find out what ordinary people think', but since, for financial reasons, they were not advertised widely in advance, few other than representatives of NGOs with an interest in GM crops and food turned up (Sample 2003).

The various experiments in involving the citizen in the decision-making processes of scientific and technological advance have therefore had mixed success. While only a few are privileged to be closely involved, they have provided rare opportunities for public debate, and helped encourage debate and openness in areas where, for commercial or other reasons, secrecy sometimes obtains (Rothman, Glasner and Adams 1996). Irwin (1995: 173) acknowledges that there is no single model or blueprint to meet the challenge of inclusive involvement in satisfactorily bringing together the social, environmental and technical issues.

Conclusion

The issues at stake in discussing relations between science, food and agriculture, the State, and the public at large, in Britain and Europe, dramatically highlight the need to recognise the contingent nature of knowledge production in late modernity. We have not attempted to map the full range of issues, insights and debates which constitute the GM controversy. However, by focusing on this important case study at the interface between humankind and the natural world, we have suggested new ways to better understand the complex social and natural processes involved in the introduction of this new technology. This understanding informs a better appreciation of the policy-making processes that assess potential risks and develop appropriate regulatory mechanisms to deal with them. It also promotes greater participation and involvement in these processes through insights into how risks are framed by experts, and the role of lay expertise in the management of new technology. We hope that it effectively illustrates, through clarifying the socially and politically co-constructed nature of social life, how developments in science and technology cannot be divorced from society at any level.

Chapter 6
Globalisation and the Transformation of Nature

The globalisation debate is both complex and unresolved (see for example Scott 1997), and here is not the place to develop it further by critically reviewing its contours. Suffice to say that many commentators have noted the demise of the traditional nation-state with a resultant stripping of many of its functions, and the development of various global entities that have taken on their mantle in key areas such as communications, culture and the economy (see inter alia Bauman 1998, Castells 1997, Giddens 1990, Hirst and Thompson 1996). The result has been the rise of what some term super-states, which are successful only because the nation-states while residual in many areas, are still strong enough to provide the sense of identity necessary for the exercise of citizenship. Into this debate, predicated, as so much of it has been, on the new communication technologies (Castells 1997), we wish to insert some thoughts about the new genetics. As already noted in earlier chapters, we are of the view that the questions it raises about the relationships between the social and natural world are likely to be at least as far reaching in their consequences as those of any previous technological advance.

In particular, the chapter, following an account of the relevance of the globalisation debate, will discuss the case of the commercialisation of the new health, food and crop technologies, including intellectual property rights and technology transfer especially between the global North and South, including the new areas of pharmaco-genetics and pharmaco-genomics. The chapter will conclude with an analysis of risk perception in the context of changing conceptions of space and time which the new technology has generated, and the implications for developing an alternative framework for viewing issues of ethics and morality.

The Regulatory Role of the Nation-State

It has been suggested that all modern states are nation-states, consisting of a political apparatus which legitimately lays claim to specific territories, able to uphold this claim through the control of military power, and encompassing citizens who feel committed to its national identity (Giddens 1989, 302-305). Following TH Marshall, citizenship is said to bestow certain rights: civil rights such as freedom of speech and action; political rights such as universal franchise; and social rights including enjoyment of minimum standards of welfare and security. Together these have, in Western societies at least, formed the basis of what has come to be called the Welfare State. A significant implication of this development is that the nation-state is able to control its own economic development. But, as Bauman (1998: 56) notes, being able to balance the books is 'becoming more and more an actuarial fiction' as the nation-state becomes eroded by the transnational forces of globalisation. While the process has been going on for some time, the disappearance of the two great power-blocks following the dissolution of the Soviet Union has forced a redefinition of 'global politics' and revealed that 'no-one seems now to be in control'. No longer have two group of states 'parcelled-out' the world. What has been revealed is the collapse of the three bases for upholding statehood – military, economic and cultural self-sufficiency – with the result that 'The very distinction between the internal and the global market ... is exceedingly difficult to maintain in any but the most narrow "territory and population-policing" sense' (Bauman 1998: 65). Hence in shedding its key foundations, the state has been performing a striptease in the cabaret of globalisation and left itself with only the bare necessities: 'its powers of repression' (Bauman 1998: 66). In economic terms, these bare necessities function to facilitate the free flow of global capital and shield the citizen from market anarchy.

The nation-state is therefore of continuing importance, and indeed, in its attenuated form, continues to proliferate as evidenced by the growth in recent years of membership of the United Nations. The breakdown of a global order based on the alliances of two opposing power blocks has also led to an aggregation of states attempting to develop their own cartels. One example is the European Union, which Castells (1997: 266 et seq) suggests is not so much a political federation as a *super* nation-state in which some degree of control of global flows of wealth, information and power can be established. We are therefore witnessing a process not of 'reinforcement of nation-states, but the systematic erosion of their power in exchange for their durability' (Castells 1997: 268). But, he argues, nation-states are not disappearing because of the growing role played by international institutions and supernational consortia. He quotes Hirst and Thompson (1996: 17) as suggesting that 'the central functions of the nation-state will become those of providing legitimacy for and ensuring the accountability of the supra-national and subnational governance mechanisms'. Hence the announcement of the death of the nation-state is premature because its policies and regulatory powers can 'ease or block movements of capital, labor, information, and commodities ...'. Multinationals are heavily dependent on *local* economic conditions and the protection of their home states for their global successes. Bauman expands this further:

> Between themselves, the states police orderly conditions in localities that increasingly have become little more than transit stations in the world-wide travel of goods and money administered by the multinational (more correctly: non-national) companies. Whatever has remained of economic management in state politics is reduced to competitive offers of attractively profitable and pleasurable conditions (low taxes, low-cost and docile labour, good interest rates, and – last but not least – pleasant pastimes for all expenses-paid travelling managers), hopefully seductive enough to tempt the touring capital to schedule a stopover and stay for a little longer than the refuelling of the aircraft demands.
>
> (Bauman 1993: 232)

The significance of this insight is illustrated by the fact that the US and Europe account between them for 90% of the practice of world biotechnology. They have been the key players (with, to a lesser extent, Japan) in the world-wide attempt to map and sequence the human genome since its inception in 1988. The Human

Genome Organisation (HUGO), dubbed the 'United Nations' for the human genome, with a brief to co-ordinate research, foster co-operation, facilitate the exchange of information, and encourage public debate on its ethical, legal, social and commercial consequences (Cook-Degan 1994) was born in Switzerland in the same year. Together with attempts to map and sequence a variety of other plants and animals, and the development of bioinformatics, these global projects have ushered in what Rifkin (1998) calls the 'biotech century'. Genes, he suggests, have become the 'green gold' of the new millennium.

In 1999, there were 1300 biotechnology-related ventures in the USA, 220 in Europe and 60 in Japan. By 2002, the USA had 1,466 public and private companies, Europe had grown to 1,878, while the Asia/Pacific region had only 108. However the market capitalisation of the biotechnology companies outside the USA was relatively small, and growing smaller. For example the capitalisation, at $52 billion, of all the European biotech publicly traded companies in 2001 was less than Amgen's $61.5 billion (Ernst and Young 2003). 63 per cent of all biotechnology drugs are being developed in North America, 25 per cent in Europe, and 12 per cent elsewhere in the world. Commercialisation of new genetics has, it appears been uncritically embraced in the United States where it is estimated that 45 per cent of all biopharmaceuticals are sold (Persidis 1998). The predicted market for biotechnology products by the year 2025 in the USA alone is US$2,520 billion (Saegusa, 1999). Since the arrival of New Labour government in Britain, there has also been a deliberate attempt to encourage local, and attract multinational, biotechnology enterprises to the UK. However, as a result of the history of food scares, and the debacle of Mad Cow Disease, the British public (and indeed the public in Europe as a whole) has been less welcoming than its American cousins (Levidow 1999).

By 2001, GM crops around the globe covered 52.6 million hectares, and four countries account for 99 per cent of the total area , with the largest being the USA at 35.7 million and Canada at 11.8 million (Soil Association 2002: 9). The commercialisation of modern biotechnology has prompted countries to look to international agreements to prevent further creation of trade barriers following concerns about safety and labelling. The World Trade Organisation (WTO) established a dispute procedure in 1995, and a number of conflicts have since been resolved on a bilateral basis. However problems with the implementation of international standards remain. While there is agreement that member states undertake risk assessments when human, animal or plant health may be threatened, there is no agreement over what constitutes 'acceptable' risk, or what is an appropriate methodology for assessing it (OECD 1999:57).

Globalisation and Polarisation

It has been acknowledged for some time that 'knowledge' is the most important global factor in determining standards of living in today's world, and the new genetics is becoming a major contributor in the twenty-first century, as the project to map the human genome comes to its conclusion. However, a number of rifts have appeared which suggest that there are forces of contra-indication in the globalisation models so far developed (Sagar et al 2000). The largest proportion of the world's population, some 80 per cent, are to be found in the developing countries of the global 'South', generating only 20 per cent of global gross domestic product. By 2003, the level of inequality between North and South had doubled, with the richest one per cent of the world's population receiving as much income as the poorest 57 per cent. The income of the richest 25 million Americans is the equivalent of that of almost 2 billion of the world's poorest people, and some 54 countries are relatively poorer than they were a decade earlier (Elliot 2003). Contributions to developing countries in the form of financial aid, from the OECD's Development Assistance Committee had actually fallen, as a percentage of their member's gross national product, from 0.33 per cent in 1992 to 0.22 per cent in 1997. The US, Europe and Japan accounted for 84.5 per cent of all R&D spending in 1994, and more than 80 per cent of patents granted in less developed countries belong to individuals or corporations located in the global North.

These difficulties are thrown into sharp relief in the context of global biodiversity, nearly 80 per cent of which is located in Asia and South America. Advanced industrial countries like those in Europe and North America while being 'gene poor' are 'gene-technology rich'.

> For example, future applications of biotechnology may increasingly depend on genetic diversity. A substantial portion of global genetic resources resides in the South, while the capabilities to commercially utilise them lie in the North. This raises concerns about appropriate benefit sharing.
>
> (Sagar et al 2000: 3)

Members of the WTO agree in principle (through the UN Convention on Biological diversity, the CBD) that the two can be traded off, so that both gene-rich and technology-rich countries gain equally from the process. In practice, the balance is uneven since the technology-rich countries invest their knowledge in Intellectual Property Rights through international patents and exploit their investments to the full. The less developed countries on the other hand, while they will have spent years protecting and developing their indigenous plants, animals and microbial species, can only sell them once, assuming that they have not already been plundered through bio-piracy. An example of this can be found in the recent patenting surrounding the member of the mahogany family in India called the neem tree (*Azadiracha indica*), the name of which ironically means 'free'. The European Patent Office, early in May 2000, revoked one patent of the fifty-one related to the products of this tree to develop the many anti-fungal agents derived from it. While this was a specific not a general patent (since an Indian company already markets the products), it proved a notable victory for the international Green coalition whose motto is 'Free the Free Tree' (Nature 2000).

There is no mechanism similar to that protecting IPRs, to enable national governments to exercise their sovereign rights to determine access to genetic resources (Kate and Laird 1999). Wheale and McNally (1998) give the example of a biodiversity prospecting contract between Merck Pharmaceuticals, the US-based multi-national, and the National Biodiversity Institute of Costa Rica in 1991. Merck agreed to pay an undisclosed percentage of any royalties, and US$1.3 million to aid conservation. However the fact that Merck's annual sales in 1991 were US$8.6 billion, rising to US$13.28 billion by 1997, compared with the Gross National Product of Costa Rica at only US$5.2 billion, rising to only US$9.5 billion in 1997, puts the deal into perspective.

These global divisions have been exacerbated as huge trans-national biotechnology corporations have developed from mergers and acquisitions at a value of US$9.3 billion in 1988 to US$ 172.4 billion in 1998 (Sagar et al 2000). Some thirty large-scale mergers including between GlaxoWellcome and SmithKlineBeecham, and Pfizer and WarnerLambert, took place between 1980 and 2000 (Walsh 2002: 157). While biotechnology is likely to continue to underpin the globalisation process, problems arise as issues of control, access and influence over agendas become increasingly polarised between North and South. The top five multinational biotechnology companies own more than 95% of gene transfer patents. The intrusion of private global capital into the South resulting from this will only serve to make the South more dependent as its choices for the indigenous promotion and expansion of biotechnology become more constrained. Vandana Shiva, a longstanding activist in this area, and one of the BBC's Reith Lecturers in May 2000, summarises the fundamental contradiction in this way:

> There is no epistemological justification for treating some germplasm as valuless and common heritage and other germplasm as a valuable commodity and private property. The distinction is not based on the nature of germplasm, but on the nature of political and economic power. (Shiva 1991: 58)

Hence any discussions of the role of the nation-state in a globalised world needs to recognise that only very few (perhaps the super-states of the global North) are able to set the agendas for the future innovation and exploitation of the new genetic technologies.

The Regulatory Role of the Nation-State

Both the US and Europe illustrate the local importance of the residual nation-state in providing profitable and congenial environments in which multi-nationals can flourish. The Food and Drug Administration (FDA) in the United States has been accused of deceiving the American public in the early 1980s into accepting that genetically modified foods were safe when the documents produced by its own scientific advisers showed otherwise (Brown 2000). Eleven of the seventeen experts whose views on potential risks were sought by the FDA expressed disquiet but, it is claimed in a legal challenge, were over-ruled. The arguments hinged on the extent to which genetically engineered plants are substantially equivalent to normal plants. If they were deemed to be so, then there would be no need to raise difficult questions about potential risk, and hence jeopardise commercial exploitation. In the event, the FDA ignored the contrary advice, and legitimated the wholesale introduction of genetically modified soya and maize in US agricultural production. Rifkin, writing before this became public knowledge, is able to point to the importance of the regulatory framework of the nation-state in providing a suitable environment to develop and attract global capital:

> For the most part, the scramble for fame and fortune corrupted the entire regulatory process, with government officials, corporate executives and molecular biologists working side by side to assure the quick and expedient introduction of genetically modified organisms into the environment, always mindful of the need to maintain the position of US eminence in the emerging new field of biotechnology. (Rifkin 1998: 79)

The US legal system also plays a key role in underpinning regulation by the state. In an article entitled 'Intellectual Propriety', *Nature Biotechnology* (2000: 469) argued that the current willingness of the US patent office to award patents for gene sequences should not be a cause for concern as any injustices could be resolved by the courts. Thus Dupont (reported on page 467 of the same issue) is currently suing Monsanto alleging it stole the gene technology associated with Round-up Ready Soya when Monsanto, acquired a subsidiary, Asgrow, in 1996, which had licensed the technology from Dupont in the early 1990s. This gave Monsanto a two-year advantage over Dupont's own genetically modified herbicide resistant soya. But, as the report notes, this is likely to be a lengthy and costly battle, and thus conveniently overlooks the unequal opportunities such recourse to the law provides for the small and less wealthy firms and social groups.

In Britain in 1990, the Government's Advisory Committee on Releases to the Environment (ACRE), predominantly made up of senior scientific experts, including some from biotechnology companies, was asked to rule on the release of genetically modified organisms (GMOs) into the atmosphere. Its brief was to develop guidelines for all applications for release, whether they were drugs, crops, foods or pesticides. The committee developed a risk assessment procedure based closely on one originally developed for the chemical industry. It also involved a protocol requiring the experts to agree an imagined future risk scenario. The outcome of this essentially rhetorical procedure, according to Jasanoff (1995), was to align biotechnology with a less novel form of hazardous activity, and deflect any difficult questions about the future risks associated with GMO release. It also ignored the fact that, as Beck (1992) and Giddens (1990) have suggested, the potential hazards associated with this new technology are *beyond* imagination. In this way, the British Government was offered, and gratefully accepted, advice that eased the regulatory framework for the development of genetically engineered plants and crops by both UK and multinational companies.

These two case studies illustrate the conclusions, which can be drawn from the current debate about globalisation. Nation-states like Britain and the USA (perhaps a super-nation state, to follow Castells) today play significant roles in the development of any key new technology by providing a congenial regulatory environment within which multinational companies in particular can flourish. The GlaxoWellcome-SmithKleinBeecham pharmaceuticals merger announced in January 2000 raises the concept of 'geo-regulatory tourism', or indeed any form of economic 'tourism', inherent in the idea of global capitalism, and used, for example, in the same way as cheap labour has been used by Nissan or General Motors. Even as this merger was being arranged, the Chairman of Glaxo-Wellcome threatened to pull much future R&D out of the UK because his new anti-influenza drug, Relenza, had not been approved for use in the national health service by the government's newly constituted regulatory agency, the National Institute for Clinical Excellence (NICE). His argument, in a leaked private letter to the British health secretary, made it plain that if the regulatory climate in Britain was perceived in foreign markets as rejecting Glaxo-Wellcome products, his company would have to consider moving abroad (Atkinson 1999). Following the merger, the new multi-national conglomerate announced that its operational headquarters would be moved to Pennsylvania, USA.

Pharmacogenetics: Of 'Snips and Chips'

The mapping and sequencing of the human genome has, it is hoped, provided a 'blueprint' of the genetic makeup of humankind. It will not, however, be able to describe the makeup of any single individual since all human beings differ, albeit only marginally, and so there will be a normal variation created by the individual's or population's distinguishing characteristics. In healthcare, these are most likely to be found in pharmacogenetics, the study of how an individual's genetic makeup influences, or is influenced by, drugs developed by the pharmaceutical industry, and pharmacogenomics, the commercial application of genomic information, derived from the Human Genome Mapping Project, to pharmacogenetics. Underpinning the growth of pharmacogenomics is the recognition that some of the variations in DNA sequences are associated with the increased risk of developing specific diseases, with variable responses to different therapies, or with adverse reactions to certain drugs. If the sequence variation (polymorphism) is present in at least one per cent of the human population it is labelled a single nucleotide polymorphism, or SNP ('snip') for short. SNPs, it is expected, will improve both diagnostic techniques, and treatments within healthcare systems that increasingly focus on value for money. Attempts are being made to patent SNPs so that drug companies can obtain suitable returns on their investment. The outcomes of this process of commodification are summarised in the expectations voiced by two founders of a major US biotech company, Variagenetics (one of whom is an academic at the Massachusetts Institute of Technology):

> Drugs that are potentially toxic will be avoided, effective therapies will be prescribed sooner, and diseases will be more effectively and economically managed.
>
> (Housman and Ledley 1998)

Others have suggested that these extrapolations, while well-meaning, are more uncertain than their champions – often combining their academic neutrality with large stakes in emergent biotech enterprises – imply, and that in any event, the time-spans involved take us far into the century. They also point to some of the associated social costs, particularly concerning the issues of privacy and discrimination (Snedden 2000). They overlook the fact that these are very much pre-occupations of the wealthier nations of the world, and that many of the diseases which form their focus, are those more likely to be found in the overweight and ageing populations of the global North, than the youthful and undernourished ones in the global South.

This is particularly true in the case of adverse drug reactions (ADRs) resulting from the normal treatment of everyday diseases. ADRs do not normally hit the headlines, although it has been estimated that in the USA some two million people suffer from them, and of these over one hundred thousand dies each year. ADRs are the fourth largest cause of death in the USA (Poste 1998). However, the activities of drug companies, and the degree to which they ignore attempts by government to regulate their activities, were highlighted by the death from an ADR of a young American patient undergoing experimental gene therapy (Campbell 1999).

These reservations have not prevented pharmaceutical companies from sinking vast amounts of venture capital into what the *Scientific American Presents* (1999) foresees as the future for health care and diagnostics – the 'snip-chip' and its associated self-diagnosis, drug dispensing technology. According to *Nature Biotechnology* (2000: 469) US$1.2 billion was raised in the first quarter of the new millennium for initial biotech offerings and another US$12.6 billion was raised by already quoted companies. However, much was lost following the Clinton-Blair statement on 14 March 2000 on the rapid release of human genome sequence data, which was followed by some personal remarks by Clinton urging that the raw data become publicly available. These together highlight a continuing debate on whether the sequence data should be made freely available which started at the 1996 Bermuda meeting of the largest public and commercial sequencers, and has remained unresolved (Marshall 2000). The commodification of knowledge resulting from the mapping and sequencing of the human genome is therefore exemplified in this aspiration to produce objects, as part of the global economic enterprise, which are far removed from the so-called 'building blocks of life'.

This process is replicated across the spectrum of new genetic technologies such as the production of genetically modified crops and foods. According to Persidis (1999), the exploitation of the 'vast and lucrative agbiotech markets' is dependent on the genomics arising in part from the effort to map and sequence the human genome. The new sequencing technologies and the gene expression assays have, until now, been used primarily in developing new drugs for healthcare use. However the smaller biotech companies find themselves in a very competitive market, and are increasingly looking to use their techniques in the agbiotech field. Large multinationals are looking for partners with specialist, niche skills. This does not mean that the difficulties in manipulating plant genomes are any the less challenging since there are similar requirements to deliver and express genes in a precise and controlled manner (Persidis 1999). The future is likely to include the use of transgenic plants to aid the reclamation of contaminated soils, to provide the basis for the factory production of vaccines, and to the targeting of specific treatments for plant diseases – a plant equivalent of pharmacogenomics. It may well be that a significant proportion of the world's needs for fuel, fibre, food and some medicines will develop from agricultural biotechnology.

This idealised future is based on a commodification process described by Goldberg (1999: BV6) as the 'business of agriceuticals':

> The farm supply firm has become a life science company. The farmer has become a custodian of land and water resources and an applier of genetic technology ... and adds value in terms of health and nutrition in a manner that reduces pollution and provides for long-term sustainability. The farmer is also expected to manage the process in a way that safeguards the diversity of germplasm and wildlife. The assembler acts as handler of identity-preserved products who ensures the origins and safety at all levels of the food chain. The processor is a developer of branded and own-label products that now provide not only calorific content but also health and nutrition alternatives. The distributor becomes the credible partner to the consumer, now working in partnership with the hospitals and health system to provide unique foods for people to manage disease control and the health and nutrition of the general population.

However, Goldberg acknowledges that this model requires a close alignment between public and private decision-making (the companies and the regulators) and the close involvement of consumer groups, if it is to succeed. There is some evidence to suggest that the contradictions inherent in the globalisation process make such success unlikely. We discuss these issues further in the next chapter.

The Global and the Local

It is probably true to say that modern industrial society is in some ways much less a dangerous place than it was during industrialisation in the nineteenth century, or even in the predominantly agricultural economies which preceded this. Those of us fortunate to live in the global North are a great deal safer (not to say wealthier) than those in the South. We also have greater expectations concerning the levels of safety we experience, greater controls over them, and ultimately greater responsibility for their management. However, Giddens (1991: 184) has noted the ubiquity of crises in modern life, defined as those occasions when activities concerned with 'important goals in the life of an individual or a collectivity suddenly become inadequate'. One reason for this is that life in a globalised world has a different meaning now than it had before.

> Everyone still continues to live a local life, and the constraints of the body ensure that all individuals, at every moment, are contextually situated in time and space. Yet the transformations of place, and the intrusion of distance into local activities, combined with the centrality of mediated experience, radically change what 'the world' actually is.
>
> (Giddens 1991: 187)

Those of us living in the North therefore experience the world very differently even in the age of risk. For example, our late modern culture is now far removed from the production process when compared with the predominantly subsistence economies of the South. Macnaghten and Urry (1998: 135) suggest: 'We are too modern for a nature we almost pass by and we are "antiquated" in front of a culture that is already passing us by.' De Waele notes, we are no longer even *homo faber*, let alone ape-man. Consequently we imagine nature as a reconstructed memory of the past rather than a lived reality. We no longer remember, for example, that 'organic' food is a human not a natural product, cultivated and engineered (admittedly without the use of contemporary genetic engineering techniques) for a purpose. Our perception is obscured by an idealisation of the past so that, as De Waele suggests, we no longer 'see', for example, real tomatoes, but only think we do.

> You rather have an image of tomatoes, of how they grow and ripen, of soil and sun and humidity and the like, than you really bodily experience. We are living with images, images 're-deconstructed' from memory, from publicity, from dreams and desires, ideas and ideals ... We are more and more distant from nature, we don't know it any more, we cherish it, we created an image of it. We live with an image of nature that is past, gone by. Nature, wilderness is shrinking and we are reconstructing for ourselves an image of nature that is in fact re-deconstructed out of a past reality and out of a never-been-like-that reality. Therefore, not so deeply inside ourselves, we want 'natural' tomatoes, the tomatoes we are imagining.
>
> (De Waele 1997: 134)

This example serves to further highlight the complexities of the globalisation process where the multinational agribusiness's, busy developing and selling GM seeds to farmers using the need to feed the poor countries of the South as justification for making fortunes in the North, stand accused of going against Nature and therefore putting the world at risk. De Waele suggests that what is at risk is a re-deconstructed view of Nature, mediated by the cultural perceptions of those living in the global North who have long forgotten what is required to survive in the 'real' world.

Linked to this perception is another aspect of globalisation, the transformation of time and space. Beck (1992) suggests, as we have noted in an earlier chapter, that science no longer performs its experiments in the bounded space of a laboratory, where the method of trial and error recognised that outcomes, while unpredictable, were at least controllable. The world has become a global laboratory, and the outcomes of the experiments carried out in it are still unpredictable, but can no longer be controlled. Adam (1998, 2000) argues that in the new genetic technologies this unpredictability has been further exacerbated by the compression of time through commodification.

> From this perspective, the genetic engineering of food is about money. It is about the promise of massive time saving in the scientific production of change. This is achieved by controlling time: controlling the seasonality of animals and plants; controlling generational sequence and reproduction……[geno-technology] has the potential of realising the time rationalizers' dream: instantaneous change in unlimited quantities, effected not at the phenotype but at the genotype. At a stroke, changes introduced in the present alter the life-course of evolution forever.
>
> (Adam 2000: 139)

Hence Nature is not only perceived differently in the North as opposed to the South, but expectations about time and space will also differ depending on which end of the commodification process is being experienced. The global market marches to a different tune, conducted by multinationals predominantly located in the super-states of North America, Europe and Japan, viewing the world as a whole as their laboratory.

Citizen or Consumer?

In this chapter, we have discussed the process of globalisation in the context of the new genetic technologies related particularly to health, and to the production of crops and food. We have suggested that while the nation-state is being stripped of many of its key foundations, it still has, in its *super* state manifestations, important roles to play in maintaining a suitably congenial environment for global capital to flourish. Our analysis of the commodification of the information derived from the new genetics points to major problems linked to contradictions in the globalisation of biotechnology. While many of these are economic, some relate to the more subtle changes which fall under the general rubric of living in a risk society. Throughout, we have highlighted the great disparity between discussion located in the wealthy countries of the globe compared with the majority who are becoming poorer to an ever-increasing extent.

The most significant conclusion to be drawn from the foregoing discussion of the relationships between the new genetic technologies and the processes of globalisation is that much of what was previously seen as nature has become transformed into forms of coded information. The social also changes through globalisation, disrupting the coherence of nation-states, and deepening the divisions between the North and the South. The distinctiveness, which marked the boundaries between humans and the world, has largely been lost as the mapping and sequencing of human and other species comes nearer to its conclusion. We are faced with recognising that 'nature' is both socially and culturally constructed, and can no longer form the basis for defining the ethical or moral foundations for a 'good' life. Perhaps, as Macnaghten and Urry (1998: 31) suggest,

> we may increasingly live not in a risk society, which implies fixities of institution and social order, but rather in an indeterminant, ambivalent and semiotic risk culture where the risks are in part generated by the declining powers of the nation-state in face of multitudinous global flows ...

The implications for understanding ethics are far-reaching. As suggested at the start of this chapter, the ideas of citizenship bound up in the welfare role of the nation-state imply a degree of moral responsibility which is undermined as the forces of globalisation begin to strip the nation-state of its foundations. Bauman (1993) argues that this has made moral responsibility something that needs now to be paid for rather than seen as embedded in the political and social institutions of the state. A significant justification for the rapid development of pharmacogenomics has been that new drug treatments will provide better value for money. But, following Bauman (1993: 244), this implies that the ideal for the citizen is to become a satisfied customer, with little regard for the costs elsewhere: it is 'your value for my money'. This is nowhere better illustrated than early attempts to develop the so-called 'terminator gene', which would ensure trouble-free crops for the farmers year-on-year, but only if they purchased new seeds from the same producer on an annual basis. Far better, then, to recognise the need to bolster the nation-states of the South (rather than attempt to sell terminator genes to Indian farmers) through more equal collaboration and generous financial assistance, so that the scientific and technical human power and infrastructure can facilitate the best local use of indigenous knowledge (Vasil 1998). We recognise that this represents a rather oversimplified account of the complexities of understanding the operation of the globalisation process in relation to the new genetics, and accept the view of the 1999 Earthscan report that:

> The richness and complexity of the legal, political, scientific and socio-economic framework for the commercial use of bio-diversity does not lend itself to generalities and simple conclusions. (Kate and Laird 1999: 315)

However, we also wish to support the 1993 UN draft Declaration on the Rights of Indigenous Peoples which it quotes as asserting the right of self-determination and the territorial and resource rights of *local* people as being critical to the effective distribution of benefits to *local* communities and to the conservation of biodiversity (Kate and Laird 1999: 313, with our emphasis).

We may have dramatically increased the stock of human knowledge, but simultaneously, we appear to have made acting on it in any moral fashion more difficult.

> Moral responsibility prompts us to care that our children are fed, clad and shod; it cannot offer us much practical advice, however, when faced with the numbing images of a depleted, desiccated and overheated planet which our children and the children of our children will inherit and have to inhabit in the direct or oblique result of our present collective unconcern. Morality which has always guided us and still guides us today has powerful, but short hands. It now needs very, very long hands indeed.
>
> (Bauman 1993: 218)

It is our contention that an analysis based on the existing institutions of morality are no longer sufficient to deal with the globalised nature of the biotechnology industry. They are unable to address the question: which patients get whose drugs? They provide only 'cottage industry' ethics to deal with 'multinational' issues such as the exploitation of the resources of the poorer countries of the South to ease and extend the quality of life of those in the North.

Chapter 7
From Commodification to Commercialisation

The commodification of genomics, the subject of this chapter, resonates with a number of issues discussed earlier; issues such as the attempt to privatise the human genome sequence (Chapter 3), the privatisation of indigenous peoples' knowledge of plants (Chapter 5), and the exclusion of large numbers of countries and people from the potential benefits of genetic knowledge (Chapter 6). All of these might be seen as a socialisation of areas of life and nature traditionally occurring in a relatively narrow or restricted frame of life to a broader sphere within which they are more amenable to the current needs of Capital. We shall examine this process with respect to genomics to show how this recently won knowledge, and much of the social structure that produced it, is commercialised through a complex and ever changing process of technological innovation; in particular the vital and controversial part played by patenting. We also present case studies of several leading genomics firms from the perspective of their evolving business models. The term 'genomics' predates the HGP, but not 'genome' which had been in circulation for many years. Thomas H. Roderick coined the neologism in 1986 for the title of a new journal (*Genomics*), defining it as the discipline concerned with

> sequencing data, discovery of new genes, gene mapping, and new genetic technologies [as well as]...the comparative aspect of genomes of various species, their evolution, and how they related to each other.
>
> (Chitty 2003)

Whilst the term is useful, for our purposes here we need to be aware that such a broad definition poses boundary problems which need to be taken into account when analysing genomics business activity.

Commodification and Science

On a number of occasions we have drawn attention to the work of Merton and his CUDOS theory of the norms of science, notably in Chapter 5. This traditional social model of science, if it ever closely corresponded to reality, is now being reconstructed by an intensification of the commodification of science seeking to draw or push scientists into the business of selling and buying knowledge. Scientists who, like John Sulston and his colleagues in the Public Consortium, were socialised into the traditional norms, and who laid great value on science as public knowledge feel put down and threatened. On the other hand, commodification has a progressive aspect for it opens the way to new solutions to human problems through the innovation process. Potentially it overcomes what Bernal (1939: 380) in an earlier era called the 'frustration of science', of course, not in the manner he envisaged. Capitalist economies, contrary to what Bernal thought, proved able to encourage and make use of science; the novel twist being that science is commodified, public knowledge is privatised. Thus there is a contradiction in that the resulting technological progress is a reflection of market needs, which as we saw in Chapter 6 do not necessarily meet everybody's needs, and creating also a 'Risk Society' (Beck 1992).

Marx observed that the wealth of capitalist societies appears as 'an immense collection of commodities' (Marx 1976: 125), and that in order to reproduce and expand itself Capital requires a stream of new sources of commodities. Commodities, he argued, have a dual character (Marx 1976: 131); they have a 'use value', that is they have a utility, and an exchange value, that is they can be bought and sold in the market place. How can genomics, or particular parts of it, be commodified, that is be turned into something which has an 'exchange value'? In the form of scientific knowledge genomics clearly has a use value to scientists in their endeavours to advance scientific understanding, but it doesn't necessarily have an exchange value, being freely donated within the scientific community. Of course, this not to say that scientific research costs nothing, as we know from the HGP it can be a very expensive activity. Nevertheless, unless it is conducted in secret in a military or private organisation, the findings of scientific research are generally publicly and freely available for use by other researchers possessing the appropriate scientific skills and research resources. When this norm is not observed or is threatened, as it was from time to time during the HGP, then as we have seen the scientific community becomes aggravated. The problem with such a purist view of scientific knowledge is that the costs of research, which were over $3 billion in the case of the HGP, require funding sources, and somewhere along the line the patrons of science expect some kind of payback. How this is obtained is a complex multiform process, and is often a source of conflict between the interested parties. A great deal of contemporary science, technology and innovation policy is concerned with finding ways of doing this. Fundamentally it requires that some scientific and technical knowledge be commodified, that is, transferred to someone through 'the medium of exchange', converted into an exchange value.

One of the great economic changes to have occurred since Marx's time is the widespread realisation of the major role that scientific knowledge can play in commodity production by expanding the forces of production. This role is not necessarily simple or direct, for as innovation research has demonstrated the traditional linear model of Basic Research-Applied Research-Market Need no longer reflects reality, if it ever did (Dodson and Rothwell 1994). No doubt it was a useful heuristic for a time, serving to focus the official and public minds on the view that scientific and socio-economic progress were linked, and so help the scientific community present 'a model for future appeals…to the public purse' (Steinmueller 1994: 55). It might be argued the originators of the HGP presented such a model of the benefits that might accrue from the project.

Current models of the innovation process stress market 'pull' as well as research 'push' operating in complex networks of institutions and knowledge flows. In the case of genomics commodification, given the recent origin of the field, we might expect a large element of research push to be involved, however, sophisticated networks rather than simple transfer from the laboratory to market place will be involved in successful commodification. The commodification of genomics comes at an interesting period in the development of those actors who will need to play important roles in innovation networks which could spawn a host of new genomic-based products and processes. The recent evolution, or rather co-evolution, of these actors, needs to be examined, albeit somewhat briefly.

Initially genomics research – and here we need to remind ourselves that its scientific boundaries overlap several established disciplines with their own histories, for example, genetics, molecular biology, evolutionary biology – was advanced in university and publicly funded laboratories. During its development, as we have seen with the HGP, several industrial sectors, financial institutions and public interest organisations paid attention to its progress, and became actors in the field themselves.

In recent years universities and other public science institutions, have undergone many changes – often painful – which enable them to be more closely connected to industry and commerce which are thus able to draw more easily on the latest research findings coming from publicly funded science. The cross-over from academia to commerce is most apparent in the US, where for instance over 300 biotechnology companies have been founded by past and present faculty members of just five Californian universities. The clustering of biotechnology start-ups around leading academic and public life science research centres in the US and Europe is further evidence of the changes we are discussing (Shorett et al 2003: 123). Henry Etzkowitz (2002) has described the phenomenon as

> An industrial penumbra appears around scientific institutions such as universities and research institutes, creating feed back loops, as well as conflicts of interest and commitment between people involved in the two spheres. Over time, as conflicts are resolved, new hybrid forms of science-based industry and research, as well as new roles such as the entrepreneurial scientist are created. New organizational models are also invented as innovations in teaching and research lay the groundwork for the entrepreneurial university.
>
> (Etzkowitz 2002)

For some such as Steven Shapin (2003: 18) these developments are a mixed blessing that might presage the emergence of a new divide in academia between '...those academic practices that have goods to sell, and commercial options, outside the academia, and those that do not'.

In the mid-nineteenth Century, when the links between science and industry were rarely consciously sought by entrepreneurs, Marx was speculating on science becoming a direct force of production (Marx 1973: 699). Since then the links between industry and science have steadily, if unevenly, developed. However, for much of that period research tended to be conducted separately in academic, governmental and industrial labs. Roughly speaking, academia specialised in basic research, whilst government labs dealt with research applied to national social and military goals. Today the situation is one in which boundary crossing and hybridisation between disciplines, and university, government and industrial R&D institutions is becoming the norm rather than the exception. It is not necessary to describe these changes in detail here, indeed their interpretation is still contested (Elzinga 2002). Gibbons et al (1994) speak of a transition from traditional scientific research, which they refer to as 'mode one research', to a new type 'mode two research', a fundamental feature of which is its closeness to industry and other sites of application. Etzkowitz and Leyersdorf (1997) propose the metaphor of a 'triple helix' of academia, government and industry. There is empirical evidence for these changes, for example, scientometric studies of US patent trends have shown US industry to be massively dependent on recent public science research. Nowhere is this more the case than the biotechnology industry which, despite being barely two decades old, is more dependent on public science than other industries; biotechnology firms showed a greater reliance on public science than even the big pharma, which are highly research intensive (McMillan et al 2000).

Human DNA Patents

Patents allow the holder a monopoly for around 20 years, the exact period varies between countries, in return for public disclosure of details of the invention. In this way society benefits and progresses from new technical understanding and the inventor benefits by having his invention protected from being stolen. To be granted a patent the inventor has to give evidence, among other details, that the invention has a use, is novel, and has an inventive step. It should not be a product of nature or a scientific discovery. In practise the rules are not easy to apply, especially by overworked and overloaded patent offices dealing with claims in new areas of technology. The details of patent law vary between countries. The US laws, for example, have proved more flexible than Europe's in obtaining patents in biotechnology, and dealing with the 'product of nature' issue and living organisms.

A legal landmark in the US, which seems to have improved the likelihood of obtaining biotechnology patents, was the Chakrabarty case in 1980 (Wade 1980). Chakrabarty and his employer, General Electric, had filed a claim for a microorganism they had developed which was able to degrade oil slicks. The claim was initially rejected because microorganisms were products of nature and living. The US Supreme Court agreed by a small majority that the microorganism had been sufficiently altered, that it was a composition of matter and not a product of nature; the fact that it was alive did not in this case prevent the patent being awarded since the key issue was that the organism was '... the result of human ingenuity and research'. Without this interpretation the development of biotechnology might have proved more difficult since most biotechnology start-ups have used their IPR to lever finance. Other countries followed the US opening the way for patenting cell lines, microorganisms and genetically manipulated organisms. There are now numerous patents dealing with genetically manipulated organisms, such as the Harvard oncomouse. The right to patent such transgenic animals, when designed as disease models, is generally accepted, despite ethical objections that the practice is cruel.

Universities in US and Europe have over the last two decades significantly increased their patenting activity, though it is small compared to corporate patenting. In some countries there have been government and local policies to encourage university patenting (Etzkowitz and Webster 1995). In the US the Bayh-Dole Act 1980 allowed universities and non-profit organisations to obtain intellectual property rights for inventions obtained through federally funded research; more than 150 US universities possess patent portfolios, and more than 2000 companies have been formed through university licenses. Biotechnology is an area in which patenting by academic institutions has been frequent. One of the fundamental biotechnology patents, describing how to genetically engineer an organism and invented by Stanley Cohen and Herbert Boyer, was owned by Stanford University and the University of California, San Francisco – which have earned over $200 million in royalties and licence fees from the patent. Columbia University which owns the Axel series of patents covering gene splicing techniques earned over $300 million during the lifetime of the patents. The ownership by universities of such blockbusting patents is, however, exceptional.

The possibility of patenting DNA sequences raised anew many of the issues which had emerged in the early days of biotechnology. Partly to test the matter the NIH in 1991 and 1992 made patent applications on over 2000 ESTs produced by Craig Venter, then an NIH employee. The claims related not only to the ESTs but also to the genes they represented and to proteins encoded by the genes. There was an acrimonious dispute in NIH over this policy, which in part was to blame for the departures from NIH of James Watson, then head of the NIH Genome Centre, Bernadine Healy, NIH's Director, and Venter. The US Patent and Trademark Office (PTO) rejected the initial patent applications, and by 1994 Harold Varmus, Healy's successor, decided to forgo pursuing the claims further. The matter of DNA's patentabilty remained unclear and throughout the 1990s the EPO began allowing patents on DNA sequences and genes, and was later followed in this by the European Patent Office and the Japanese Patent Office. The matter is still not settled. As late as 2003 the European Commission is threatening eight EU members with legal action because they have not come into line with an EU Directive, 98/44/EC, which has sought to bring European biopatent law more into line with US law. This directive allows the patenting of 'an element isolated from the human body or otherwise produced by means of a technical process', and also states 'the sequence or partial sequence of a gene may constitute a patentable invention, even if the structure of that element is identical to that of a natural element' (Halbeck 2003: 960).

An analysis of DNA patents by Thomas et al (2000: 1185) found that 18,174 patents had been filed by the end of 2001. During the period 1996–1999 62% were from US organisations, 20% from Europe and 10% from Japan. The number of patents filed by public-sector organisations grew in the period, accounting for a third of the patents.

Critics of current patenting practice maintain that DNA sequences are discoveries not inventions; whereas patent offices have taken view that sequences are artificially isolated not natural genes. Some countries such France and Germany object to patenting of human materials and genes on the grounds that they instrumentalise the human body. The question of patenting in this area is fraught with difficulties which have unfortunately created a climate more favourable to the legal profession than to science. Clearly there seems no simple route out of this quagmire; the European

Commission is being pressured by researchers to halt the awarding of broad patents on DNA sequences by making the criteria of inventiveness and utility more stringent (Mitchell 2002). One argument for tightening up the criteria for granting DNA patents is that advances in technology have made gene identification much simpler. Maybe the early DNA sequences and genes found through positional cloning techniques had a sufficient degree of inventiveness to merit patents, but this is no longer the case with the contemporary *in silico* methods of identifying genes from gene databases (Nuffield Council on Bioethics 2002).

A report from the Nuffield Council on Bioethics summarised concerns about the growing numbers of DNA patents:

- the cost of research may increase, as the grant of increasing numbers of patents will mean ever more licences are required before research can be conducted
- research may, as a matter of practice be made more difficult if researchers are required first to negotiate the use of patented genes and sequences
- a patent holder may withhold a licence to gain maximum financial benefits, or licence it exclusively to one or a limited number of licencees
- companies that wish to acquire the rights to several DNA sequences may decide not to develop a therapeutic protein or diagnostic test because of the number of royalty payments that would be required (this is sometimes referred to as royalty stacking)

(Nuffield Council on Bioethics 2002: 59)

Another worry is that ready availability of DNA patents will lead to too much litigation. There has been an increasing number of contested biotechnology patents; the resulting litigation is now averaging in the US $1–3 million per case. Often this proves a waste of money that might have been better spent on R&D for human benefit; for some biotechnology firms litigation costs have led to loss of useful technology or even bankruptcy (Perelman 2002).

In Chapter 3 we described how the leaders of the HGP fought against attempts to patent the human genome and ensure that it remained freely available. Will academic research now be frustrated further down the line by DNA patents, or will it be protected by the 'research exemption' practice? Most patent systems operate a 'research exemption' allowing academic non-commercial research to make free use of patented inventions. The aim of this is to ensure that the patent system does not stifle research. Unfortunately, the exact boundaries of the exemption are hazy and can be contested, especially where, as in the US, research exemption is not based on a statute. A recent US court ruling has threatened their research exemption practice leading to fears that academic research could be slowed down and made more expensive (Malakoff 2003). The Nuffield Council on Bioethics (2002) recommended that Europe clarifies the practice of research exemption, and that US makes it statutory.

Emerging Genomics Firms

The development of biotechnology has been closely intertwined with the pharmaceuticals industry, which has faced diminishing returns from R&D in recent years. Fewer 'block buster' drugs are reaching the market, whilst the R&D costs per drug have increased. Drug development is a risky business because of the high level of scientific, technical and clinical uncertainty involved. Biotechnology offered the possibility of a new paradigm for drug discovery, one which was more rational and science-based. Genomics can be regarded as the latest element in this process, thus many of the current commercial opportunities for genomics-based firms are associated with the needs of the pharmaceutical industry. The scientific findings and technologies of genomics can be applied to various stages of drug discovery such as target identification and validation, and assay development. Big pharma are looking to integrate genomics into their drug discovery programmes, usually not in-house since it seems to be more convenient at this stage to do so in collaboration with specialised genomics firms. Early genomics firms often had specialised sequencing expertise which they used to generate genetic databases, which big pharma were able to subscribe to through collaborative deals. Later, developments in proteomics, pharmacogenomics, functional genomics, and bioinformatics opened up new opportunities for commercialising genomics. Exactly how many genomics firms there are is not clear, to some extent the number depends on how one defines a genomics firm. For example, Cook-Deegan et al (2000: 13) defined a genomics firm as one in which 'a substantial fraction or all of their business plan hinges on use of large data sets containing information about DNA structure, or depend on its interpretation'. Based on this they identified 161 genomics firms, of which 64 were publicly traded. For many of these genomics was only a part of their business.

Genomics Business Models

Tate (2000: 4) defines the notion of business model as

> … shorthand for what a company intends to be its *winning formula*. Combining focus, strategy, philosophy and core expertise, the term describes a vital and multifaceted frame of reference within which the company plans to conduct its daily business, add value, take charge of its destiny, and compete successfully in its chosen market.

The business model should integrate a firm's external relations such as products, markets and customers, and internal ones such as organisational structures, core competencies, and finance; it describes the organisation and structures adopted by a firm for making a profit.

To understand the emerging genomics business models we need first to examine those in biotechnology. Biotechnology is not strictly speaking an industry, rather a grouping of start-up firms, or new biotechnology firms (NBFs), built around a generic technology based on genetic engineering and molecular biology able to provide novel products and services to established industrial sectors such as pharmaceuticals and agriculture (Orsenigo 1989). A wide range of business models have been tried in biotechnology. Some involved relatively simple applications of genetic engineering, producing small volume products not requiring expensive efforts to meet safely regulations. Another approach was to provide biotechnology research services. Most models allowed the NBFs to occupy niche positions in relation to established firms in pharmaceuticals and agribusiness, often involving various forms of collaboration. There was no ideal biotechnology business model in a world of constant change, what worked in the early days might fail later, and models which were not initially viable might become so. There was, however, one business model which many NBFs aspired to and few succeeded with; this was a fully integrated pharmaceutical company (FIPCO), for example Genentech. This business model involved drug development, from research and development to market. The model has a high revenue potential, for instance AmGen in 1999 had revenues of $3.34 billion. There are, however, significant barriers to entry, including patent protection costs, high development costs and marketing costs. But, as in the case of British Biotechnology, failure can be disastrously expensive. Most of the drugs produced by NBTs, termed biopharmaceuticals, are based on recombinant forms of natural proteins, natural product and monoclonal antibodies. After a slow lead period of nearly two decades the rate of production of new biopharmaceutical seems to be increasing. Up to the end of 1999 only 84 were on the market in US and Europe but during the period January 2000 to June 2003 a further 60 had been approved for clinical use, and about 500 more were in clinical evaluation (Walsh 2003: 865). Few of the biopharmaceuticals, so far, have turned out to be big money makers. Many of the biopharmaceuticals are now developed by the big pharma such as Schering-Plough or Eli Lilly, but several biotechnology firms which adopted the FIPCO model still figure strongly, for example Genentech, Biogen, Chiron and Genzyme.

During the 1990s new technological advances and research associated with the HGP created new opportunities for a new wave of biotechnology entrepreneurs especially through tool based/technology platform models. These are able more rapidly to generate revenue streams than the FIPCO model through licensing fees, milestone payments, royalties, and service fees from big pharma and biotechnology companies.

The genomics business is rapidly evolving and its boundaries are fuzzy, and any current analysis is soon likely to be overtaken by events; firms change their strategies and business models, some develop hybrid models, and others may be taken over or become bankrupt. Our aim here is not to produce a definitive survey of genomics firms, but rather to examine how genomics is being or might be commercialised. Various classifications might be made of the business models used by genomics firms, the one we have adopted here presents four types (Biospace 2000):

- Structural genomics
- Functional genomics
- Target drug discovery
- Enabling genomic technology

For each of these we have presented case studies, five of the six case studies are chosen from the leading group of US genomics companies listed in Table 7.1.

Table 7.1 Leading US Genomics Companies (Source: Dooley 2002)

Company	Market Capitalisation, 2000($m)	Sales, 2000 ($m)
Millennium Pharmaceuticals	5,885	196
Human Genome Sciences	5,374	22
Celera Genomics	1,556	43
Incyte Genomics	1,042	194
Affymetrix	1,032	201

Structural Genomics Companies: Examples, Incyte and Celera

This business model was used by genomics information companies working on the structure of the human genome, and its related proteins. Initially companies such as Incyte and Celera built databases of genomic sequences, SNPs, and functional data on genes and proteins; they sought to make money by selling subscriptions to the databases. The basic difficulty with the business model is that information in such databases can soon become obsolete, and the business strategy has to take this into account if the business is to remain sustainable.

Incyte Genomics (Incyte 2003)

Incyte Genomics was founded in the early 1990s on a strategic vision that a health care market existed for genomic information and genomic technologies. Incyte pioneered the genomic information business several years before the formation of Celera. Since 1991 it has systematically isolated and characterised ESTs, placing them in its own proprietary database. It has focused on the discovery and analysis of genes, especially those expressed in disease transcribed genes, and was one of the first to patent genes. For subscription fees researchers could search the Incyte database looking for new drug targets, and generally advancing research on a range of biomedical issues. Initially there was some scepticism over a business model of non-exclusive database subscriptions. However, Incyte has so far managed to survive without the need to produce pharmaceutical products of its own. It does, however, own one of the largest collections of patents on human genes, over 400 in 2000 and growing, which may have a high potential market value.

Over time Incyte has adjusted its business model to changing circumstances, as data which had originally been unique to its proprietary databases became available in public databases after the completion of the human genome sequence, and in those of rival firms, such as Celera. Incyte, like other genomics firms, strengthened its business through R&D collaboration and partnerships, and by widening the range of its proprietary offerings. Its current technology platform includes genomic and proteomic database products, genomic data management software tools, microarrays, and related reagents and services which Incyte provides to the pharmaceutical and biotechnology industries to assist their drug discovery and development efforts.

The range of Incyte's corporate collaborations and nature of its technology platform can be seen from Table 7.2. The collaborations support the R&D activities of the clients, some of whom also collaborate in advancing Incyte's own R&D.

Table 7.2 Incyte Collaborations (Incyte 2003)

Agilent Technologies, Inc.	Agilent licenses the rights to use Incyte's database; high-value sequence-verified, non-redundant cDNA clone collections; and gene patent portfolio for the commercialisation of microarrays.
Amersham Biosciences	Incyte has an agreement with Amersham Biosciences to offer Incyte's pre-assembled sets of human and mouse clones in a format customised for Amersham's suite of integrated, gene expression microarray products.
Epoch Biosciences	Epoch Biosciences partners with Incyte to make oligonucleotide and other hybridisation probes using Incyte's proprietary sequence content.
Galapagos Genomics NV	Incyte and Galapagos work together to conduct functional genomics studies using Galapagos' cell-based assay programme and pharmaceutically relevant full-length gene transcripts from Incyte's database to help annotate and prioritise drug targets.
Iconix Pharmaceuticals, Inc.	Incyte and Iconix work together to develop and commercialise information to provide chemists, pharmacologists and toxicologists in the pharmaceutical and biotechnology industries with an integrated, flexible resource to understand and predict the potential mechanisms of action, toxicity and other biological effects of candidate drug molecules.

Lexicon Genetics, Inc.	Incyte and Lexicon Genetics use the technology and intellectual property assets of both companies to accelerate the development of therapeutic protein drug products.
Linden Bioscience	Linden Bioscience licenses the rights to use Incyte's linear RNA amplification technology to make and sell linear RNA amplification kits for internal research use that leverage Linden's proprietary solid phase transcription chain reaction technology.
LION Biosciences AG	Incyte licenses LION's SRS integration platform to provide Incyte's subscribers with a user-friendly Web interface.
Medarex, Inc.	Incyte and Medarex work together to develop human antibody therapeutics. Medarex and Incyte will share the cost and responsibility of preclinical and clinical development of antibody products and will jointly commercialise any antibody products resulting from this collaboration.
Odyssey Pharmaceuticals, Inc.	Incyte and Odyssey work together on functional characterisation of proteins encoded by Incyte-proprietary gene transcripts.
PerkinElmer Life Sciences	PerkinElmer Life Sciences has access to Incyte's comprehensive collection of proprietary cDNA clones from Incyte's database for the commercialisation of microarrays.

Celera Genomics (Celera 2003)

The origins of Celera have already been described in Chapter 3 where we described its launch by PE Systems and Craig Venter, and the challenge it posed to the HGP Public Consortium's programme to sequence the human genome. Here we shall look at its business model and how it has changed and evolved since the company's foundation in May 1998. Celera Genomics is one of three business segments of Applera Corporation. Applied Biosystems and Celera Diagnostics (a 50/50 joint venture of Celera Genomics and Applied Biosystems) are the other two segments.

Celera Genomics was formed as a bioinformatics and discovery company using its technology to analyse, sequence and map the genome. Venter described Celera as an information company whose business model was analogous to those of Lexis-Nexis, Bloomberg and AOL (Venter 2000). Celera was the only commercial company to directly sequence the human genome, others such as Incyte went along the cDNA route.

The business model initially offered access to its database of the human genome, clearly such an offering had relatively short commercial life as the human genome became a public property through the HGP. Celera began to annotate its data to make it more valuable and unique. It soon had several big pharma as subscribers to the database; these included Novartis, Pharmacia and Amgen. Subscribers had, according to what they paid for, access to databases, annotation and tools. An initial aim of Celera was to identify and patent around 300 human genes; it initially filed provisional patent applications for 6,500, from which it would pick the 300 genes, after appropriate studies and consultations with partners.

When Celera went public it raised nearly $1 billion, then the largest ever biotechnology IPO. At its peak its share price rose to $247 in February 2000, during the hype generated around the completing of the sequencing of Chromosome 22 and the Drosophila genome, and by Celera's race against the Public Consortium. In nine months the shares had gained over 3,000 per cent. Three years later they were trading at about $10. There were several reasons for the fall. Celera shared the general loss in confidence in high technology investments after the dotcom bubble burst. Then there was a slowdown in the US economy. Finally, and more seriously, investors began to lose confidence in the Celera business model and doubt whether it could bring financial returns commensurate with its share value; by 2002 its stock market value was little more than its $940 million cash holding.

The company was forced to rethink its business model and announced early in 2001 a strategic shift to drug discovery. Its declared mission now

> … is to discover and develop meaningful new therapies that improve human health. We are applying our diverse genomics, bioinformatics, proteomics, medicinal chemistry and biology technology platforms to identify and validate drug targets, and to discover novel therapeutics.
>
> (Celera 2003)

As a result of these changes Venter resigned from the company in January 2002. It is too soon to say whether Celera will successfully make the transition to the FIPCO model.

Functional Genomics: Example, CuraGen

This business model is based on the ability to determine functional data on genes. Patents are generally used to protect proprietary information. The main markets are human health, and agriculture. Typically this model seeks deals with customers, for example, large pharmaceutical companies, which sell data for development fees, and royalties on products reaching the market. The main challenge involved in adopting this business model is protecting IPR and staying competitive until a drug development strategy is possible, which is thought by investors to be more sustainable.

CuraGen (Source CuraGen 2003)

CuraGen is a US firm which was founded during the period of the HGP. It went public in 1998, with a slogan of 'Delivering on the promise of genomics' by understanding gene and protein function and their role in disease, and the goal of becoming a leader in functional genomics. The technological basis of its business is shown in Table 7.3.

Table 7.3 CuraGen Technology Platform

- Bioinformatics
- Gene sequencing and SNP discovery
- Expression analysis, gene discovery, and pharmacogenomics
- Genotyping, disease gene discovery and pharmacogenomics
- Expression confirmation real-time quantitative PCR, and *in situ* hybridisation
- *In vitro* assays for each disease programme
- *In vivo* animal models, including transgenic, knockout models and gene trapping

As CuraGen has developed, it has begun to move beyond functional genomics and seek longer-term revenues and sustainability through creating a portfolio of pharmaceutical products, using collaboration and alliances with Abenex and Bayer. At the same time it has continued service-based collaborations. Table 7.4 shows the nature of a selection of its collaborations and alliances.

Table 7.4 CuraGen Collaborations and Alliances

ALLIANCE	TOPIC AREA
Abinex	Antibody drugs
Alexion Pharmaceuticals	Oncology drug target and validation
Bayer	Obesity and diabetes codevelopment, pharmacogenomics and toxicogenomics collaboration
Biogen	Drug target discovery and pharmocogenomics
COR Therapeutics (now merged with Millennium Pharmaceuticals)	Drug target discovery and pharmacogenomics

DuPont /Pioneer Hi-Bred International	Agricultural product discovery
Genentech	Drug target discovery and pharmacogenomics
GlaxoSmithKline	Drug target discovery and pharmacogenomics
Hoffmann-LaRoche	Drug target discovery, pharmacogenomics and diagnostics
Mitsubishi	Central Nervous System/Schizophrenia Drug Target Discovery
Monsanto	Agricultural Gene Discovery
Ono Pharmaceuticals	Pharmacogenomics and Toxicogenomics
Pfizer	Pharmacogenomics, Toxicogenomics and Proteomics
Roche Vitamins	Animal Health
Sequenom	Drug Target Validation/Human Genetics Novel Target Identification and Validation through Proteomics

CuraGen's business model might now be seen as a hybrid, in which building on its technology platform and experience servicing drug companies and through appropriate alliances, it moves towards a genomics pharmaceutical model. It now has a pipeline of protein, antibody, and small molecule drugs in the areas of obesity and diabetes, oncology, inflammation, and central nervous system disorders.

Target Drug Discovery: Examples, Human Genome Sciences, Millennium Pharmaceuticals

This model is a bold one which from commencement adopts a strategy of seeking to develop genomics-derived drugs. Very few firms have adopted it, although others have evolved towards it through other business models. These firms aim to become fully fledged pharmaceutical companies able to discover, develop and market new drugs. The major challenge for the model is the clinical testing phase, which proved the downfall of some of the first generation biotechnology companies which tried this route.

Human Genome Sciences (Human Genome Sciences 2003)

Human Genome Sciences, Inc., was founded in 1992 to pioneer genomics and human gene discovery for developing pharmaceutical products. It was the first genomics firm to aim to establish a drug development business model; it currently defines itself as:

> a biopharmaceutical company with the mission to discover, develop, manufacture and market new gene- and protein-based drugs. HGS is a pioneer in genomics, the systematic study of all genes of an organism, and a pioneer in converting genomic knowledge into drugs to treat and cure disease. HGS is dedicated to victory over disease.
>
> (Human Genome Sciences 2003)

The company was started by a former Harvard University professor William A. Haseltine, a charismatic individual, notorious for media spats with rivals such as Craig Venter. Ironically it was through an alliance with Venter that HGS was established. After Venter left the NHS in 1992 a venture capitalist, Wallace Steinberg of HealthCare Investment Corporation, offered to commercialise Venter's EST work. Venter was initially reluctant to join a commercial company because he felt that it would impede his ability to publish. Steinberg then brokered a deal whereby two companies were established. One was TIGR, a not-for-profit research centre which eventually received funding of $85 million from Steinberg; TIGR later developed into the major centre for shotgun sequencing of genomes. The second was a for-profit company, Human Genome Sciences, which was given privileged rights to commercialise TIGR's research. Steinberg introduced Venter to Haseltine, who had acted as consultant to the venture capitalist over the development of start-up biotechnology companies. Haseltine quickly saw the scientific and commercial implications of the EST work, and was appointed CEO of HGS. Haseltine and Venter later fell out and the two organisations went their own way (Davies 2001).

In 1993 HGS signed 'the consortium agreement' with SmithKline Beecham (now GlaxoSmithKline), at the time this was a pioneering development. The agreement was later modified in 1995 to include Takeda Chemicals Ltd, and in 1996 Schering-Plough, Merck KgaA, and Synthelabo (now Sanofi-Synthelabo). HGS sold access to its EST database to members of the consortium, which hoped to use them as a starting point for new drugs. Later other big drug companies copied this approach, but realising drugs has not proved easy. Although genetic databases yielded many potential targets understanding how bad genes cause disease proved more difficult. Nearly a decade elapsed before GlaxoSmithKline began clinical testing of its first genomics-based drugs. HGS' agreements with its consortium partners remained co-exclusive until July 2001, since then HGS has possessed exclusive access to its technology for its own developments, except for any genes which are subject to ongoing research and clinical development programmes by consortium members.

HGS began a functional genomics programme in 1997; an integrated programme brought together all HGS' technologies. These included; gene discovery, bioinformatics, molecular biology, cell biology, pharmacology, protein chemistry, high-throughput biological screening, medical and regulatory affairs, drug formulation, manufacturing and strategic marketing. HGS always adopted a strong patent strategy to protect the genetic and protein structure and functional information in its databases. It believed the firm's structural functional approach allowed much stronger IPR claims than would be the case if it relied on claims based on DNA sequences encoding proteins of unknown function. It also reserved downstream rights on the use of its patented proteins, and sought royalties on drugs developed by subscriber companies using its data. HGS currently is entitled to milestones and royalties from the drugs resulting from 430 research programmes, involving 300 target genes, being pursued by consortium partners. HGS estimated that the consortium members' research programmes include less than one per cent of the genes discovered by HGS, and that the majority of drug targets identified by HGS remain as candidates that it might develop on its own or with other partners. HGS has built a manufacturing and process development plant to develop and manufacture gene and protein-based drugs, and now has several drugs undergoing clinical trials.

Millennium (Millennium 2003)

Millennium is a US firm founded in 1993; it successfully built upon an integrated technology platform based on genetics, genomics, throughput screening, and bioinformatics to move to a drug discovery business model. Originally its technology platform was used in drug-target discovery, but it was able to move towards a drug development strategy through a series of strategic alliances and mergers and acquisitions, see Table 7.5 and Table 7.6.

Table 7.5 A Selection of Millennium's Strategic Alliances

YEAR	COMPANY	AGREEMENT
1994	Roche	Contract licensing gene targets, royalty
1995	Lilly	Contract licensing gene targets, royalty
1998	Bayer	Discover 225 genomics-based drug targets, potential value $465 million
2001	Aventis	Discovery and development small molecules for inflammation, value $450 million
2001	Abbot	Co-development and co-commercialisation metabolic treatments for obesity and diabetes
2001	Affymetrix	Co-development GeneChip applications for drug discovery and development
2002	Ingenuity	Knowledge management systems

Table 7.6 Millennium: Mergers and Acquisitions

Year	Company	Expertise and Products
1999	Leukosite	Campath, Lymphatic leukaemia
2000	Cambridge Discovery Chemistry	Chemistry
2002	COR Therapeutics	Integrilin, cardiovascular diseases

Millennium developed a very successful strategy for its move into drug development by using its partnerships for funding and broadening its capabilities. Its strategic alliances with leading pharmaceutical and biotechnology companies had by 2003 generated almost $2 billion and its early deals with Roche and other big pharmaceutical companies also enabled it to use its technology platform to leverage expertise in product development and marketing. This has enabled Millennium to move down the value chain seeking more control over products; by 2001 it was able to make co-development and co-commercialisation deals which allowed for 50/50 ownership of any new products. Today it is a pharmaceutical company with two products in the market and others in clinical testing.

Enabling Genomic Technology Firms: Example, Affymetrix

This business model is based on the supply of instrumentalities. The development of genomics instrumentalities was described in Chapter 2; the business is about supplying new research tools, sequencers, reagents, diagnostic tools, hardware, and software etc. for doing genomics and related research. The challenge for this model is remaining at the forefront of innovation and marketing of the technology; obsolescence and substitution are ever present dangers.

Affymetrix (Affymetrix 2003)

Affymetrix is the leading DNA chip technology firm, its technology is used for acquiring, analysing, and managing complex genetic information for use in biomedical research. The company applies the principles of semiconductor technology to the life sciences to create gene chips that are of great value to genomics research.

Affymetrix's GeneChip technology was invented in the late 1980's by a team of scientists led by Stephen Fodor. The technology was based on a revolutionary idea that united semiconductor manufacturing techniques with advances in combinatorial chemistry to build vast amounts of biological data on a small glass chip; the first prototype was produced in 1989. This technology provided the basis for a new company, Affymetrix, formed as a division of Affymax, N.V. in 1991, and operating as an independent firm from 1992, with its headquarters in Santa Clara, California.

Affymetrix's chip and array technology have become standard tools for analysing complex genetic information; used by pharmaceutical, bio-technology, agrochemical, diagnostics and consumer products companies, as well as academic, government and other non-profit research institutes, to analyse the relationship between genes and human health.

The key Affymetrix product is the GeneChip System, the components of which include; disposable DNA probe arrays containing genetic information on a chip housed in a cartridge, special reagents, equipment for introducing the test sample into the probes, a scanner to read fluorescent images from the probe arrays, and specialised software to control the system and also to analyse and manage genetic information. This system is designed for large business, academic and clinical labs; smaller array systems are made for individual researchers.

The business strategy of Affymetrix has been to seek to maintain technical leadership in the DNA probe market. This has forced it to obtain access to complementary technologies and resources via a variety of acquisitions, agreements and collaborations. For example, in order to maintain and improve its product Affymetrix has made acquisitions, including; Genetic MicroSystems which specialised in DNA instrument technology, and Neomorphic, a private computational company developing bioinformatic tools. Some of the collaborative agreements are listed in Table 7.7.

Table 7.7 Affymetrix Collaborations (Affymetrix 2003)

COMPANY	AGREEMENT
Agilent Technologies	Scanner supply
Beckman Coulter	OEM supply gene chip arrays for disease management products
BioMerieux Vitek	Collaborative agreement focusing on bacteriology and virology disease management products; industrial and food testing
BMS, Millennium Pharmaceuticals, Whitehead Institute	Collaborative agreement in functional genomics and polymorphism discovery
Enzo Diagnostics	Agreement labelling kit supply
Genetic Analysis Technology Consortium	Agreement to establish standards for DNA array-based products
Mouse Genome Consortium	Collaborative agreement to supply funding for determination of the sequence of the mouse genome
Orchid BioSciences	Collaborative agreement to develop genotyping assays
Roche Molecular Systems	Collaborative agreement to develop disease management arrays

We have illustrated our discussion with successful firms, but there have been several notable failures of genomics-based firms. One such is DoubleTwist, a US bioinformatics firm. It was founded as Panagea Systems in 1993 as a software developer. It changed its business model and name in 1999 and raised $76 million in venture capital. It represented a particular business model which did not make its own genomic data but used specialised software to search and reassemble public genomic databases to produce gene data which could then be sold to clients via a web portal. DoubleTwist announced it would, in collaboration with Sun Microsystems, compete with Celera by analysing the public version of the human genome for a fee. Despite claiming to have identified 65,000 genes, and another 40,000 possible genes it withdrew its IPO in late 2000 and closed down shortly after (Frederickson 2003).

Genset was another major failure of a public genomics company. It had based its business strategy on seeking drug targets in the fields of the central nervous system and metabolic disorders and moving on to developing its own drugs. It had before failing discovered and started developing a drug of its own.

Such examples show that there is no guaranteed success in commercialising genomics. The bioinformatics business model, in particular, seems to have been a difficult one to sustain; we have described how giants such as Celera and Incyte soon abandoned it and moved on to drug development models. No doubt there will be niches for the bioinformatics firms but by 2001 the venture capital markets preferred product business models to bioinformatic platform models. The reasons for the difficulties with the bioinformatics platform model were that there were too many start-up firms seeking to sell high-priced products to businesses drawn from a relatively small pool of big pharmaceutical firms, which were finding it easy to bring bioinformatics in-house.

Characteristics of Genomics Firms

A survey by Cook-Deegan et al (2000) on the funding for genomics research contains a number of important observations with implications for the future of the commercialisation of genomics. These are:

- Private funding of genomics has overtaken public funding
- US dominates genomics start-ups
- US dominates genomics patents
- Commercial genomics ignores the Third World
- Genomics is dominated by speculative investment

Cook-Deegan et al (2000: 8) calculated that in 2000 public and non-profit funding of genomics was about $820 million compared to 1999 figures of $845 million for 29 publicly traded genomics firms. To this might be added a similar figure for spending by biotechnology and pharmaceutical firms. These were all approximate estimates but the authors were confident that private sector annual spending on genomics was now ‘in the range of twice government and non-profit spending’.

In Cook-Deegan et al's sample, 76 per cent of publicly traded genomics firms and 71 per cent of privately held genomics firms were based in the US, further, much of the non-US genomics private research is carried out in the US and genomics-related patents are concentrated in the US During the period 1996–1999 the US filed 62 per cent of DNA patents, the EU filed 20 per cent and Japan 10 per cent (Thomas et al 2002: 1186). It seems clear, given the dominance of US genomics firms, the US will hold first mover advantages plus much of the tacit knowledge need for commercial success.

An examination of corporate genomics research and product pipelines shows that it is heavily devoted to the health needs of the developed world, even more so than is the case with the pharmaceuticals industry generally. Of course, given the concentration of research in the US, EU, and Japan it will be very hard for the Third World to take advantage of genomics; China may prove an exception. Ethnic Chinese are working in US genomics firms and academic centres, and, if China continues its recent economic growth rates, some of these people may be attracted to start genomics businesses in China.

The commercialisation of genomics has at times been dominated by speculative investment in which spectacular rises in the share prices of the leading companies has been followed by equally spectacular crashes. Table 7.8 shows a chart of Celera's share prices during 2000. The shares peaked as we have already said in February 2000, there was a price collapse after the Clinton-Blair statement on the need for public ownership of the human genome. There was a rally around the time of the announcement of the draft sequence, thereafter, as we have already noted, the share price collapsed steadily to around $10 in 2003. This curve is typical of the other five genomics firms we have written about, and is not dissimilar to the overall NASDAQ Biotechnology Index over the same time period.

Table 7.8 Celera Genomics Share Price January – December 2000 (NASDAQ)

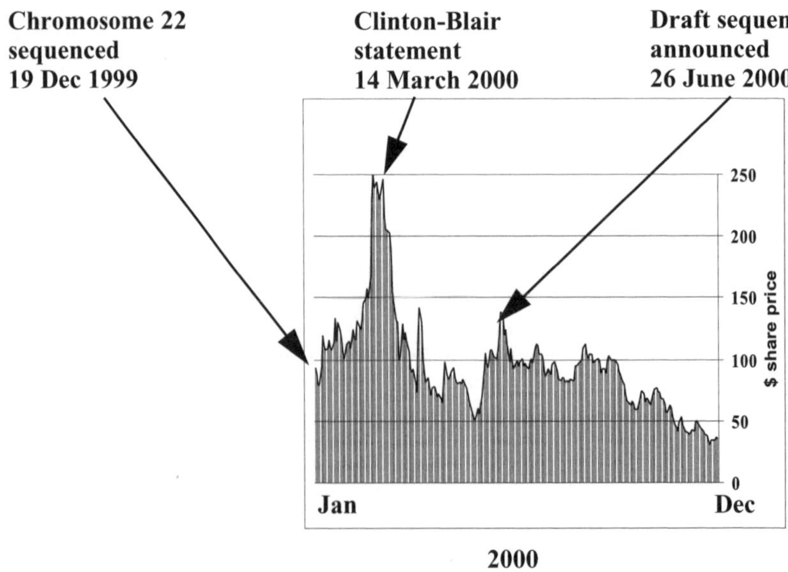

Commenting on what was seen as an over abundance of new genomics company flotations, an editorial in *Nature Biotechnology* (2001: 597) wrote

> Cynics might conclude that the multiple genomics floatations and secondary offerings of 2000 were merely a way of extracting more money from the investor community during its temporary lapse into enthusiasm…

Since 2000 the major stock markets have collapsed, especially technology stocks, including genomics firms. It is not clear what the impact will be on the development of genomics businesses. There is some evidence that investors are distinguishing between biotechnology and the computing and electronics sector, and that biotechnology has not seen a collapse on the scale of the dotcom sector. US venture capital funding of biotechnology seems cyclical, having risen from $800 million in 1999 to $2,872 million in 2000 and falling to $2,160 million in 2001; despite recent falls these investments are large sums and seem to indicate that some venture capitalists still have faith in the sector (Bradford 2003: 983). Rasnick (2003: 355), however, notes that biotechnology was making massive losses, more than $5.3 billion by US public biotechnology companies in 2001, and that in 2001 biotechnology IPOs were only raising a fraction of what they managed in 2000. We are not yet able to ascertain whether this represents a short-term phenomenon or something more profound.

Conclusions

We have argued that the development of genomic firms offers more support for the interactive theories of innovation process whereby the process moves in a non-linear fashion between science and technology and between industry and the public sector.

It also poses a warning that simplistic top-down government stimulation policies may be a waste of money, and that the benefits associated with fundamental research may not necessarily be accrued to the nation in which the research was carried out. Despite much of the HGP being carried out outside the US most of the commercialisation is concentrated there. It might be that successful policies will take account of emerging geographical clusters of genomics research, for example in the UK the 'Genome Fen' around Cambridge.

The issues surrounding the patenting of human DNA sequences and genes are proving highly contentious. Whilst the case can be made that patent protection encourages commercial development of scientific advances, there is also some evidence that some patents are frustrating scientific research.

We have raised the issue of how long existing genomics business models will remain viable. Some commentators now argue that there are now too many genomics companies, 'each holding one or two pieces of the gene-based discovery jigsaw puzzle' (Anon. 2001: 597), for many of them to become profitable. One may expect to see more collaborations, mergers and acquisitions between genome firms themselves such as the merger of Gemini Genomics (specialising in population genomics) and Sequenom (mass spectrometry-based genetic analysis) (Anon. 2001).

We have argued that the dominant 'technology platform' model may evolve via such mergers into multiple technology platforms able to tackle a larger part of the drug discovery chain, whose products can be marketed to big pharma, whose creativity may be buttressed by outsourcing the discovery stages of R&D to such firms. Further, that some genomics firms like HGS and Millennium Pharmaceuticals will turn to the 'ultimate model' – the integrated pharmaceutical firm, by using their technology platform to develop their own products for diagnosis or therapy (Papadopolos, 2000).

We now know that gene sequences in themselves proved insufficient for most sustainable commercial purposes. The real value would be found in proteins, hence the emergence of a new hype about a proteomics programme.

> This ambitious project to map the human proteome within three years will enable us to make a great leap forward in our understanding of the causes of human disease… the human proteome has become the next frontier of modern biology

enthused Peter Meldrum CEO of Myriad (Meek 2001). In the concluding chapter we explore and discuss this new post-genomics world with its talk of a 'new paradigm' and numerous 'omics', of which proteomics is but one.

Chapter 8

Rights or Rituals: Involving the People

The debate concerning the introduction of genetically modified crops and foods discussed in Chapter 5 provides a current example of the growing involvement of a significant new group of actors, 'citizens' or the 'public' in the decision-making process. In this chapter we discuss some of the new ways in which participation and involvement have been developed, beginning with a brief discussion of the move to the production of socially robust knowledge. This is followed by an in depth case study of one of these, the citizens' jury, and the chapter concludes by exploring the implications of involvement in the wider socio-political context.

Public questioning of techno-scientific advances was framed, as famously formulated by the Royal Society (1985) phrase 'the public understanding of science', by equating understanding with support and appreciation, and seeing resistance as misunderstanding or ignorance (Wynne 1995, Irwin 1995). Recent surveys have shown, however, that greater knowledge of science and technology (especially biotechnology) among the public appears (in Europe at least) to lead to greater concern (Bauer and Gaskell 2002: Chapter 12; Hampel and Renn 2000). Members of the public, as citizens, require greater involvement in techno-scientific decision-making largely because science has become demystified, and its authority de-legitimised (Beck 1992, Giddens 1990). This has posed a two-fold challenge for decision-making institutions such as science and government:

> how to manage effectively and efficiently public policy and decision issues with inherent scientific-technical uncertainties, potential risks and social ambivalence; and how to achieve public legitimacy when faced with dissenting social factions and an apparent growing public distrust in government.
>
> (Joss and Brownlea 1999: 322).

It has been suggested that the practice of science is undergoing a shift from producing *reliable* to producing *socially robust* knowledge (Nowotny, Scott and Gibbons 2001). This necessitates not only ensuring robustness through conventional discipline-bound norms but also being sensitive to a much wider range of social implications. Expanding on this, Nowotny et al (2001: 199) suggest that while scientific knowledge is always incomplete (either in the Popperian or the Kuhnian sense), it is also sharply contested. However, the boundaries within which this contestation takes place have now become extended beyond a narrow peer group and entered into the social world of what they term the 'agora' – a new public space for the contextualisation of knowledge. They describe the view of scientists as the main actors in the narrative of scientific advances as a naive 'modernist' account that ignores their role in a multi-authored story. Science is, after all, only a method believed to be superior to others only so long as it produces results.

> If the much-vaunted epistemological core of science turns out to be empty, let it be empty. What replaced this core is more interesting – shifting local practices, consisting of heterogeneous collections of beliefs, inextricably intertwined with the materiality of science, with methods and procedures, local practices which place different values on objectivity, proof and verification and combine them in different ways. These local practices are more or less robust and reliable depending on context; they comprise a medley of theoretical, instrumental and experimental elements; and they are wrapped up in a social epistemology that equips science to move at ease in the *agora*.
>
> (Nowotny, Scott and Gibbons 2001: 233)

In their view the Human Genome Project (HGP) illustrated how the culture of biology has become radically transformed along these lines. Research problems were tackled at the same time as production problems, and exposed to public anxiety and concern through the associated programmes on HGP's ethical, legal and social implications. They cite the co-operative research network of sequence production centres that produced the publicly funded first draft of the human genome as an example of the new organisational forms that will spread to break down entrenched institutional and disciplinary boundaries. They suggest that scientists themselves should recognise what social scientists have already revealed (Jasanoff et al 1995) that scientific work is co-constructed and contextualised. Progress is no longer a linear development, but arises 'in the messy and convoluted interstices that link scientific activities with the many local and social contexts with which they interact' (p. 235). The *agora* therefore facilitates greater involvement by non-experts in techno-scientific decision-making as an integral part of producing socially robust knowledge – the addition of the context of implication to the traditional context of application. While the extent to which the HGP's associated programmes on its ethical, legal and social implications contributed to developing socially robust knowledge, understanding the part played by growing public involvement in, for example the GM foods or the gene therapy debates underpins the importance of recognising its situated and co-constructed nature.

A variety of new decision-making structures, developed over the last 50 years, have been harnessed by government to facilitate the public policy-making process through an open and public dialogue in science and technology. The House of Lords in its *Science and Society* Report (House of Lords 2000) had called for a 'change in institutional terms of reference and procedures to open them up to more substantial influence and effective inputs from diverse groups'. The rationale was both to improve the quality of the decisions taken, and to ameliorate the 'democratic deficit' that was perceived to exist. The Parliamentary Office of Science and Technology reviewed a variety of innovative consultation methods in addition to more conventional ones such as questionnaires, opinion polls and invitations for written submissions. These included deliberative opinion polls, citizens' juries and panels, standing consultative panels, consensus conferences, Internet dialogues and focus groups (POST 2001). They are characterised by being *deliberative*, with participants engaging in considered debate with the possibility of modifying views, and *inclusive*, attempting to seek out the views of all likely to be affected by a decision, especially previously excluded or hard to reach groups. Unsurprisingly, the extent to which these criteria were to be found working in practice varied. One interesting example, that raises in various forms most of the difficulties of institutionalising the agora for certifying socially robust knowledge, will be explored in depth – the citizens' jury.

Citizens' Juries

Citizens' juries have been developed in both Europe and the USA as a means to improving public involvement in policy decision-making particularly in the area of local government (Cochrane 1996, Seargeant and Steele 1998). The model was introduced into Britain by the Institute of Local Government Studies at Birmingham University and the Institute for Public Policy Research in 1994, and pilot projects have been run by the Local Government Management Board, on local government decision-making, and the IPPR and the King's Fund, on health issues (Coote and Lenaghan 1997, McIver 1997). It has been taken up by the UK government as part of its approach to finding a 'third way' (Giddens 1998) in contemporary political life (see for example the use of a citizens' panel on the future of agriculture and the food system as part of the Citizen Foresight exercise, LCGIS 1998).

The success of the pilot projects in furthering the democratisation of policy decision-making is summarised by Coote and Lenaghan (1997: 88–90). They note the willingness of ordinary citizens to participate; their ability to deal with quite complex issues, arguments and data; and their developing scepticism about the quality of the evidence being presented to them. They recognise that the jury model does not easily lend itself to addressing detailed planning questions or considering abstract ideas, and that most jurors behave as though acting on behalf of the community rather than just themselves. They conclude that while citizens' juries are only one of many inputs into the policy-making process, they may, through confidence building, encourage more active citizenship: 'The great strength of citizens' juries is the opportunity they provide for informed deliberation.' However, recent research suggests that this may be overstating the case (Dunkerley and Glasner 1998, Glasner and Dunkerley 1999, Pickard 1998).

A citizens' jury is usually commissioned to address an important, local policy or planning issue by a body which is committed to act on its recommendations (Stewart et al 1994). In establishing a citizens' jury, an independent steering committee sets the question to be addressed and oversees the process. The jury normally consists of 12 to 16 jurors selected randomly to represent the appropriate constituency, who, following a detailed briefing, meet for four days to scrutinise information, cross examine witnesses, and discuss the issues either together or in small groups. They agree a report, which need not be unanimous or binding, which is submitted to the commissioning body. However, the recommendations are made widely available and the commissioning body is required to either act or explain why it has not done so. Observers, who are not allowed to interact with the jury, are encouraged, to broaden public participation and to contribute to the transparency and legitimacy of the proceedings.

The Welsh Case Study

The Welsh Citizens' Jury, organised by the Welsh Institute for Health and Social Care (Iredale and Longley 1997), was held in Cardiff in November 1997 and addressed the question:

> What conditions should be fulfilled before genetic testing for people susceptible to common diseases becomes available on the NHS?

It was an attempt to extend and develop the model in a number of ways (Dunkerley and Glasner 1998). Firstly, its members were chosen to represent a much larger population, the Principality of Wales, than had been the case on previous occasions in the UK when they were selected from local authority or similar constituencies. Secondly, the commissioning body was a large trans-national pharmaceutical corporation with no commitment to act on any recommendations the jury may provide, but with commercial interests in the topic. Instead, a list of detailed recommendations (WIHSC 1997) was submitted by the jurors in person to the Advisory Commission on Human Genetics in London. Thirdly, the focus on genetic testing required the jury to be briefed on the medical, scientific and technical background of genetic testing, in addition to the structure of the NHS, and the mechanics of jury procedure. However, both the objectivity of this process and how it may have interacted with the resources brought to it by individual jurors, are open to question.

Together, these may have contributed to mitigating the value of this approach to involving the public in the decision-making process (Glasner and Dunkerley 1999). In more general terms, wider issues such as the role of the jury as an additional constituency in a pluralistic decision-making process of health policy formation appears not to have been given sufficient thought. The fact that key actors may establish juries for consultative and educational purposes, or as part of a sophisticated public relations exercise, has also not been sufficiently explored. Finally, the social construction of a lay public which has been pre-configured by a definition of jury membership raises important questions of social control disguised as democratic emancipation. The jury approach may have become a 'technology of legitimation' (Harrison and Mort 1998), or 'subjectifying technology' (Hill and Michael 1998), that sits uneasily between active citizenship and disinterested evaluation of evidence.

The Jury in Operation

The Role of 'Local' Knowledge

One of the key issues which is said to underpin the democratising credentials of the citizens' jury approach to decision-making is the input from lay members of the public. The proponents of the process in the policy arena see the jurors themselves as providing the lay input when they evaluate the 'evidence' in arriving at their recommendations. But, there is also a growing recognition that the knowledge brought to the process by the jurors themselves cannot be overlooked. Any juror, when faced with expert opinion, does not evaluate the knowledge claims in isolation from his or her experiences and perceptions. Kerr et al (1998: 52), based on their research with focus groups, argue that:

> People are all experts about their own lives. And as social actors we engage with a range of other actors and institutions and therefore develop a unique set of knowledge from which to judge new experiences. Expertise is therefore not solely the province of professionals, but lay people have valuable knowledge and understanding of the social world which equips them to discuss the new genetics in a sophisticated and reflexive manner.

Arriving at a set of recommendations may constitute more a process of social deconstruction and renegotiation of knowledge claims than a competent or incompetent evaluation of expertise as implied by the concept of 'judging on the evidence' (see the discussion on forensic evidence in Lynch and Jasanoff 1998). Little or nothing was known, for example, about the resources brought to the Welsh Citizens' Jury by the jurors themselves, except when they chose to share these with each other during the event.

Terms of Engagement

The case study in Wales suggests that the configuration of the categories of 'lay' jurors and 'expert' witnesses contributes to diminishing the effectiveness of the jury model. There is a growing literature about the dangers of ignoring the power of experts to set agendas (Douglas 1985), define boundaries of discourse (Carter 1994), and impose assessments of risks and hazards (Wynne 1995, 1996). Kerr et al (1998: 54–55) unwittingly confirm this when they note the problematic nature of mounting any challenge to professional expertise, including the medical profession, in, for example, clinical contexts (Strong 1983). Purdue (1995: 171) in his discussion of consensus conferences also identifies the importance of what he describes as the 'terms of engagement' in defining the roles of lay and expert contributors. This conceptual weakness in the Citizens' Jury approach suggests it may be rooted in an unexplicated, and unrecognised, deficit model of expertise (Wynne 1995).

One important element in the process centres on the 'gate-keeping' role of the Steering Committee in deciding what preliminary information is made available to jurors. This concern has been echoed by Fixdal (1997) with reference to the Norwegian experience of consensus conferences. He notes the importance of the background information used to provide the lay members with a balanced account of the scientific and technical knowledge needed to discuss the topic. Much of this is written by actors in the public debate, and it is normally impractical to include all aspects of an issue. Selection by interested parties, he suggests, is inevitable. This is particularly so with the selection of witnesses by the Steering Committee in Wales, which signally omitted to provide any witness from an ethnic minority (for example sufferers from thalassaemia), or those opposed to genetic testing.

Pseudo-expertise

These difficulties may be further complicated by the fact that the jury is often presented with pseudo-scientific speculation rather than evidence-based knowledge so that the halo-effect of certain kinds of expertise can be seen to justify a wide range of responses to jurors' questions, many of which lie well outside the witness' areas of specialism. Discussing this phenomenon in the context of the National Consensus Conference on Plant Biotechnology, Purdue (1995: 171) notes that: 'Being granted "expert" status at the conference empowered one to speak on almost any aspect of biotechnology, transferring one's authority from discipline to discipline at will.' This also appeared to have happened in the Welsh case, with one medical expert often including opinions (for example about the organisation of the NHS) far removed from his specialism of paediatric medicine. While legitimately based upon his personal experience of hospital and NHS Trust work, they did not reflect the primary reason for his participation, and their contingent nature may not have been clear to the jurors.

Contingent and Empiricist Repertoire

What constitutes a scientific knowledge claim is context-specific; a conference paper is not a refereed article in *Nature*, which in turn is not a presentation to a lay audience. We make this point clearly in Chapter 5 in our discussion of Dr Pusztai and the GM potatoes. Scientists often vary the meaning of apparently similar concepts depending on their understanding of the context in which they articulate them. As we noted, Gilbert and Mulkay (1984) describe these as 'contingent' and 'empiricist' repertoires. In the empiricist repertoire, the significance of the scientist's own actions in constructing knowledge is downplayed. In the contingent repertoire, actions 'are no longer depicted as generic responses to the realities of the natural world but as the activities and judgements of specific individuals...' (Gilbert and Mulkay 1984:57). In the Welsh Citizens' Jury, it was clear that the scientific and medical witnesses mainly utilised the empiricist repertoire, and downplayed the role of human agency in the production of knowledge. The contingent repertoire was most consistently used by the General Practitioners, patients and health service experts. Interestingly, the commercial representatives and the social scientist seemed to mix the two. However, it was unclear to the evaluators whether the significance of this distinction for assessing the value of expert testimony was as evident to the Welsh jurors as evidence elsewhere suggests (Fixdal 1997, Kerr et al 1998).

Symbolic Baggage

Juries have long played a significant role in legitimating state authority, through imposing a visible check on political and judicial tyranny (Edmund and Mercer 1997). The jury represents the public both in the sense of serving its interests, and in ensuring that it comprehends the issues through a clear and accessible exposition of the case. The jury process is controlled by the legal profession through judicial intervention and legal practice, for example through the use of the adversarial approach, or concerning the admissibility of only certain kinds of evidence. This takes place in the context of a body of laws handed down from generation to generation, and democratically enhanced through recognised parliamentary procedures. Much debate among lawyers reflects the difficulties involved in their interpretation and application.

One important way in which the law provides an antidote to procedural imprecision and inconsistency is through the fiction that truth is there to be discovered in the search for justice; that there exists a neutral reality. In this respect the rhetoric of the law is similar to the rhetoric of science, since both lay claim to generate rational knowledge under strictly objective conditions, as public justification for their authority (Wynne 1989: 46 et seq). They fear that a lack of consistency will be interpreted by the public as a lack of rigour, and thus engender a loss of confidence in any outcome. In law, the institutional response is to emphasise ritual, and make the jury a significant symbolic token in the legal process.

The use of the jury model to widen the participation of the lay public in the political decision-making process – to ameliorate the so-called democratic deficit – therefore brings with it significant symbolic baggage. Significantly, the model has no 'body of laws' against which to measure its outcomes. Its value appears, primarily, to lie in ensuring that 'ordinary folk', representing the public at large, can be seen to engage fully in key events. But it also implies the transfer of the long-standing democratising credentials, and weighty legitimating procedures, of a court of law, to the messy arena of policy formation.

The Jury and 'Rituals of Precision'

The experience in Wales, perhaps because of the change in scale as compared with previous British examples, suggests that this transfer from law to policy formation is not without its difficulties, particularly in the important 'rituals of precision' (Smith and Wynne 1989, 2), which contribute so much to the legitimation process. In particular, these concern the emphasis on procedures: the segregation of the jury from witnesses and public, the interrogation of witnesses, the serving of a subpoena on new witnesses, the use of expert evidence, the process of decision-making about a verdict, and delivering the outcome and any subsequent action. These, along with other similarities between legal and citizens' juries, provide the necessary *ritual* elements in establishing the legitimacy of the outcomes, be they legal verdicts or political decisions. In the Citizens' Jury in Wales, most of these procedures inadequately mimicked those found in a court of law. The jury members often lunched with witnesses, and were able, when going out for a break between sessions, to interact with observers. Witnesses were asked to make a presentation before answering questions, rather than being closely interrogated about specific issues. The jury did wish to see witnesses who had not been asked to attend by the Steering Committee (particularly a representative of any major religious denomination), but for practical, organisational reasons were unsuccessful in their attempt. Experts were called but some were accorded greater status than others, and one was asked to both introduce and conclude the event. The moderator orchestrated the discussions of the jury to encourage a high degree of consensus about the outcome. The Recommendations were drafted by the organisers, based upon the jury members' discussions, and published after their agreement as to wording.

What are also missing in the Citizens' Jury approach compared with a jury in a court of law are the other players: the judge, the accused, and the accusers. It may be possible to argue that at least some aspects of the judicial role, particularly in maintaining order and ensuring equality of treatment for all participants, is taken on by the moderator or facilitator. However, such a person is not empowered to 'pass sentence' through an input into the final Recommendations which replace the verdict in the Citizens' Jury. The accused and the accusers are nowhere to be found since the question to be addressed must always be sufficiently neutral to allow an open debate with input from all sides. The sponsors, while known to all the participants, must be seen to stand aside from the process until the end, when they agree to act on the Recommendations. This at least has the symbolic effect of being 'sentenced'. Unfortunately, in the Welsh case, the sponsors had not entered into any such undertaking prior to the event, and were therefore not bound by the jury's Recommendations.

Representativeness and Typicality

A key aspect of any transfer of democratising credentials to the policy-making arena depends on the extent to which a jury drawn from a constituency bounded by the geographical region known as the Principality of Wales can be described as representing its views. McIver (1997,72) notes that at least three different meanings of representativeness appear in the policy literature: '... someone who is elected by means of a democratic process to represent the views or interests of those who elected him/her ... a sample of people who have been statistically selected to contain characteristics of the total population...[and] someone [who] is typical of others who share similar experiences'. She is unclear to what extent the experience of IPPR and the King's Fund in running pilot juries justifies describing them as representative on any of these criteria. At best, they can be described as 'typical in a general sense in that they might represent a typical mix of those found in a local population'. In any event, they are unlike the 'twelve good persons and true' that make up the jury in a court of law, although they should be 'ordinary citizens with no particular axe to grind' (Coote and Leneghan 1997,9).

A market research organisation was employed by WIHSC to choose the Welsh jury in an attempt to ensure the necessary independence from the sponsors and organisers required to establish the integrity of the process. It developed a multi-stage methodology which could be considered fairly robust in principle. However, it transpired that only one juror had experienced post-18 full-time education, seven had left school at the minimum age, relatively few claimed to be in full-time employment, none wished to be considered as native Welsh speakers, one was a Welsh resident but not a British citizen, and none appeared to come from one of the many, well-established ethnic minority groups. Much of this can be attributed to the necessary self-selection element in any voluntary process (unlike in a court of law) requiring a significant contribution of time (four consecutive days, although payment was made for attendance) on behalf of the jurors. As a result, it was never likely that the Welsh jury would be representative of the Principality, even in some very loose sense of typicality, thereby largely eroding one of its key democratising principles.

Representativeness and Difference

Early in the debates concerning the development of new technologies, the view that important questions involving political choices were in danger of becoming obscured by the intrusion of technical criteria fuelled the participatory impulse. Where risks were perceived as diffuse, or interests hard to define, the controversies were left either to the experts themselves or to professional activists. According to Nelkin (1982: 279), recombinant DNA research in the 1970s fell into this latter category. The transformation of 'genetic engineering' into the 'new genetics' has, however, alerted many important groups in society – women, the disabled and ethnic minorities – about both its positive, and increasingly negative, implications. With it has also come the recognition that, in late modernity, apparently global issues may have significant personal impacts, catapulting society into what is described by Beck (1995) as 'Barbarism modernised: the eugenic age'. A substantial element of this transformation rests on threats posed to the boundary between the self and the world. Such a threat is radically different from, for example, any posed by the new electronic technologies, which only temporarily liberate individuals from their particular social and biological characteristics (Bloomfield and Verdubakis 1995).

Therefore, one key aspect of representativeness which appears to be missing from discussions of the development and application of the jury process is the need to give weight to gender, disability and ethnicity. Women's experiential understandings of the issues are particularly heightened in relation to the new genetics, since many of its applications relate to genetic testing and the reproductive process (Rothman 1986). Similar comparisons can also be made with disability, whether impairment stems from illness, accident or genetic inheritance (Oliver 1996), and the two come together in the increasing tendency to terminate pregnancy on the grounds of foetal handicap (Hubbard 1986, McNally 1995). Ethnic minorities with specific single gene inherited disorders such as thalassaemia or sickle cell anaemia are also very much more focused on the issues (Bradby 1996). The Welsh jury, while able to bring some of these resources to their deliberations, only briefly addressed these issues, strengthening the argument made by Coote and Lenaghan (1997: 91) for suggesting that the importance of 'representativeness' in this case may be more symbolic than real.

This discussion of a Welsh case study of the application of the Citizens' Jury model has suggested that, in spite of increasing public participation in the policy process, the jury approach may effectively disenfranchise the citizen through being predicated upon an unarticulated model of expertise. Further it has suggested that transposing the jury model from the courts may only serve to highlight the jury's ritual and symbolic nature, especially in its attempts to represent the views of its constituency. However, a later attempt to build on the experiences of such conventional approaches to the citizens' jury has focused specifically on those who do not normally have a voice in the decision-making process, and effectively encouraged them to design their own citizen's jury. It was held in Newcastle in

September 2002, following an earlier meeting with older people's groups all over Tyneside, and focused on the topic chosen by the participants – how should new medical technologies be designed and regulated such that the lives of older people are improved, not merely lengthened? (PEALS 2003) As the Project co-ordinator notes, this bottom-up 'do-it-yourself' model appears 'to offer a method of action-research that has high potential for methodological transparency, participatory deliberation and subsequent citizen advocacy' (Wakeford 2002: 4).

However, even this model highlights the danger, noted by Irwin (1995) in his discussion of environmentalism, of artificially separating the 'scientist' from the 'citizen'. In the jury model, it appears that the public is still seated at the 'ringside' of the decision-making process. One reason for this can be traced to the complex and pluralistic structure of British health policy decision-making that forces managerial legitimacy to maintain itself by constructing the public as just another interest group. User involvement becomes a *technology of legitimation*. In Chapter 5, we discuss how this might happen in another area of concern arising from biotechnology, through the unsuccessful public debates concerning GM crops carried out in 2003 in the UK (Sample 2003). It can also become a token in the armoury of more powerful champions (in this case the National Health Service or the multinational pharmaceutical company) translated as 'playing the user card' (Harrison and Mort 1998). This suggests that an important role for juries may be educational and consultative (Dunkerely and Glasner 1998, Pickard 1998) rather than the promotion of active citizenship.

A second reason for finding the public at the ringside of the policy process has been suggested by Hill and Michael (1998). They follow Rose (1996) in arguing that society has developed a 'plethora' of ways to translate the goals of those in authority into the apparent choices and commitments of individuals. They focus on the use of surveys, particularly the Eurobarometer (Marlier 1992), to involve the public in decision-making, and argue that they depend upon the acceptance by individuals of a socially constructed 'lay person' for whom the questions are framed. As a result, they become *subjectifying technologies* acting as a form of social control, and in the case of biotechnology, a form of legitimation for commercial interests. It would appear that a similar argument could be made for the construction of the citizens' jury, especially given the symbolic involvement of a partial model of the legal process. The jury is given a role that sits uneasily in the space between participative democracy and a disinterested evaluation of expert testimony.

This being so, then it suggests a hint of pre-modernity (Irwin 1995: 133) underpinning a separation of between citizen and expert which effectively undermines the legitimating thrust of developing this model to address the democratic deficit. Also, through its symbolic tokenism and rituals of precision, the jury approach appears to sit rather too comfortably within the existing relations of production – government regulatory authorities, multinational pharmaceutical and biotech companies, and the health services – while giving the appearance of developing a critical challenging perspective. It would therefore appear that the citizens' jury model falls some way short of developing a truly reflexive approach to the transformation of late modernity. It still falls some way short of an ideal model for institutionalising the *agora* for the production of socially robust knowledge.

Chapter 9

New Genetics, New Millennium, New Society?

In this concluding chapter we try to summarise the arguments and insights developed in earlier chapters that recognise that the new genetics is at once a global, multi-national multi-billion dollar enterprise, and a very private and personal focus for concern. We recognise that, 50 years after the discovery of the double helical structure of DNA, and in the context of the completion of the mapping of the human genome, the genomic era is now a reality. We therefore look to explore the ways in which its associated complex and multi-faceted social processes are likely to unfold in a variety of policy arenas, including health, commerce, state regulation, and the law. We begin by exploring the nature of the vision of the potential that the new genetics promises, and continue by assessing the extent to which a paradigm shift in modern biology has indeed occurred. We conclude by discussing an outline of the future – a 'blueprint' for the genomic era – published after the human genome map was completed by the US National Human Genome Research Institute, and the UK Government White Paper on realising the potential of genetics for health.

In earlier chapters we discussed how the international programme to sequence and map the human genome was begun in the mid-1980s, and completed in April 2003. The result had, prior to its completion, been described variously as the 'book of life', the 'moon shot' of genetics, or the 'holy grail' of biological science (Rothman 1994, Nelkin and Lindee 1995). Its success is attributable to the work of project teams in both the public and private sectors. In the public sector, a consortium of 20 groups in the UK, the USA, France, Germany, Japan and China, made their results freely and publicly available. The rival group, Celera Genomics, restricted the use of its results within commercial limitations. The huge amount of resulting information is stored in specially designed databases that can only be accessed using customised software. Combined with the huge sums of money ready to be invested in medical applications by multi-national pharmaceutical companies, this appears to have transformed genomics into 'big science' characterised by an exponential growth in scope, complexity and resources. This transformation has been aligned with what some commentators see as a conceptual paradigm shift from linear to functional approaches to understanding the new genetics. Together these changes suggest a more sophisticated social science research agenda than the visions we discuss allow, one that characterises the new research system as an emergent, co-constructed, and situated phenomenon.

'Black Boxing' the Map of the Human Genome

The Human Genome Project began, as we outline in Chapter 2, as a scientific dream of blue-sky research into the essence of life itself. It was made possible by the introduction of techniques such as mapping and sequencing that transformed it, for some at least, into an apparently much more mundane activity, no longer revealing 'new knowledge' as for example defined by a PhD dissertation (Keating, Limoges and Cambrosio 1999: 138). These techniques were equally available for exploitation by industry and commerce as they were by academia (Rothman 1998: 283). Mapping and sequencing technology, in much the same way as genetic engineering in the 1970s, became, as we suggest in Chapter 3, the 'techno-science' of the new genetics. Balmer (1993: 89 et seq, 1996) has argued that the Human Genome Project (HGP), at least in the UK, was packaged in such a way as to align the social worlds (Fujimura 1987) of science, funding and government to create a 'boundary object' (Star and Griesemer 1989) allowing techno-scientific work to proceed. It made it possible for scientists, administrators and government to identify as the same object, the entity labelled the UK HGP, while recognising that it also represented something different within each of their social worlds. It allowed those with different views about its scientific merit to be enrolled within the HGP enterprise. Much of the work of enrolling allies (Latour 1987) and translating their interests was in the hands of leading 'spokespersons' such as Walter Bodmer and Sydney Brenner, who achieved substantial project funding at a time of financial stringency. Scientific work was accomplished by constructing a 'do-able problem' (Clarke and Fujimura 1992) allowing the UK to keep up with peers in the US and elsewhere.

A major rationale for proceeding with the HGP was its future application to improving health and the quality of life. Little or nothing was said about the commodification of knowledge, and its impact on the 'gene business' which we introduce in Chapter 5 and discuss more extensively in Chapter 7. The publication of the first maps of the human genome have seen the preliminary attempts at the 'stabilisation' of this 'technological ensemble' (Bijker and Law 1992) of social (including political economic and cultural) and techno-scientific (maps, sequences and techniques) elements. Closure in technology involves the stabilisation of an artefact and the 'disappearance' of problems. To close a technological 'controversy', one need not *solve* the problems in the common sense of that word. The key point is whether the relevant social groups *see* the problem as being solved (Pinch and Bijker 1987: 44).

However, with the connivance of both human actors, including the users of health care, the regulatory authorities, scientist, industry, and the state (not to mention Messrs. Clinton and Blair), as well as non-human ones including the genome maps themselves (as given 'voice' in *Science* and *Nature*, both in print and on CD-ROM), society seems to be rapidly transforming the results of HGP into a 'black box', seen by Latour (1987) as an obligatory rite of passage through which all now pass not only without questioning its contents, but also without even looking inside. The shock of discovering that the genome only comprises of some 24,000 genes, and not the 100,000-plus originally projected, making the differences between humankind and the rest of the animal kingdom much less significant than originally thought, has meant a radical rethink about the functions of genes as expressed in proteins. In effect the limitations posed by this discovery, and incomplete nature of the maps of the human genome, is serving to hide the contents of the book of life from public view.

The accomplishment of this process of 'black-boxing' appears to be through the simple device of admitting that molecular biologists knew all along that HGP would not provide access to the Holy Grail. For example, Paabo (2001: 1219) notes, in the introduction to the special section of *Science* (in which the map is printed) devoted to future directions for human genome research: 'Perhaps for the pragmatic biologist, the determination of the human genome sequence is a prosaic event – the delivery of a wonderful tool, but a tool nonetheless.' However, *The Guardian* (12 February 2001, on its front page) reminds its readers that the publicity surrounding the announcement of the completion of the mapping process in June 2000 had likened it to nothing less than the invention of the wheel. One longstanding critic, Richard Lewontin, in an article in *The Sunday Times* (8 July 2001, 4), seems to be clearer about the implications, under the headline: 'They got the wrong key of life'. Devotees of HGP, he notes, kept assuring the world that when we had all the genes, we would know all the proteins since genes make proteins. But '..., they now say that, of course, they knew all along that genes don't make proteins'.

All of these changes, following the mapping of the human genome and other species, are still in their infancy. Some will need new experimental strategies and technologies (Fields 2001: 1224). Together they provide a new starting point for the understanding of basic genetic make-up, with major challenges and opportunities in the fields of health and other applications. In effect, they constitute a new paradigm shift (Peltonen and McKusick 2001, Peltonen 2001) in biomedical and related research. The result has seen the creation of numerous 'omics' such as transcriptomics, metabolomics and proteomics to explore gene function in a large scale manner using what some now describe as systems biology (Blackstock and Mann 2001: S1). In the next section we discuss the evidence for describing these processes as constituting a new paradigm in biotechnology.

Is Genomics a New Paradigm?

There are a number of indictors to suggest that a paradigm shift may indeed be occurring in biotechnology. Firstly, as we suggest in Chapter 4, it is fast becoming, to use de Solla Price's phrase, a 'big science' (Price 1963). Ziman (1984: 138) characterises this transformation in relation to high-energy physics. Big science requires large and costly facilities, regulatory and policy changes, huge teams of scientists and technicians with a proliferation of specialised roles, and with the results, in the end, 'published as just one primary paper, with a hundred co-authors each seeking some degree of personal recognition' for their contribution to knowledge. This latter point is well illustrated in the publication of the competed maps of the human genome that contained the names of some 520 scientists split relatively evenly between the public and private enterprises. The public consortium of 20 groups, however, was spread across 48 laboratories worldwide indicating the complexity and magnitude of the enterprise. The fact that 'discoveries in biotechnology are no longer made by individuals, but as big science, in a collective of scientists, machines and technicians' (Keating, Limoges and Cambrosio 1999: 138), raises a number of interesting questions of organisation and management. It also points to the importance of the new communication technologies in maintaining the integrity of the research enterprise and the community of science.

Biotechnology has also become transformed, as we discuss variously in Chapters 3, 4, 5 and 7, from a largely academic pursuit to a multi-billion-dollar commercial and scientific industry (Collier 2001: 9). Blackstock and Mann (2001) go so far as to suggest that the missing '-omics' in modern biotechnology, given how capital intensive systems biology has become, is 'economics'. Elzinga and Jamison point out, these changes are even more far reaching. Biology is increasingly replacing physics as the privileged scientific area in the corridors of power and, as we saw in Chapters 5 and 8, the main source of imagery in the broader public discourse over science and technology policies and institutions (Elzinga and Jamison 1994: 596).

Secondly, following the black boxing of HGP, there now appears to be a change in emphasis away from structural genomics. The established linear, monocasual and deterministic models that have reigned in molecular biology over past decades (discussed in chapter two) have become largely redundant (Huang 2000, Eisenberg et al 2000). No gene operates in a vacuum. As Zweiger notes (2001: 126), the gene deterministic model is simply no longer sufficient to explain all that needs explaining: 'It has been stretched beyond its capacity to explain biological phenomena.' As Barbara Katz Rothman (1995: 6) succinctly put it: 'The exactness of diagnosis does not translate into any exactness of prognosis ....' Zweiger goes further to suggest that successfully addressing these anomalies will indicate the emergence of a new (Kuhnian) paradigm, although the process has only just begun (Zweiger 2001: 133–134).

The mapping and sequencing of the genome of human beings, and other forms of life (such as worms and plants) has resulted in pharmaceutical and agriceutical companies essentially using the same toolbox of functional genomics, bioinformatics, combinatorial chemistry, and high-throughput screening in a search to exploit and develop bigger and better drugs or crops. We discuss this along with the development of Genetic Data Banking briefly in Chapter 4. In a special *Insight* issue, *Nature* (*Nature Insight* 2000) describes the completion of the mapping of the human genome as the start of the post-graduate era of 'functional genomics'. Scientists are 'scrambling' to develop new techniques that *exploit* genome data to *ask* entirely new questions (Vucmirovic and Tighman 2000: 820, emphasis added).

The new 'mindset' moves away from simply cataloguing genetic information to understanding how the components work together (Lockhart and Winzeler 2000: 827). In tandem with this conceptual shift, the new genetics, as we discuss in Chapter 4, has witnessed a transformation from being relatively data-poor to being data-rich (Vucmirovic and Tighman 2000), resulting in part from new and parallel developments in bio-informatics, imaging techniques, magnetic resonance spectroscopy, and X-ray crystallography. Prodigious amounts of data (Fields 2001: 1221) are being produced and stored in large scale databanks as we note in Chapter 4, where the quantity appears to be more than doubling every year. 'We are swimming in a rapidly rising sea of data' says one observer (Roos 2001: 1260), 'how do we keep from drowning?'. Data now drive innovation through encouraging 'fishing expeditions' where questions rather than hypotheses legitimate new research (Lockhart and Winzeler 2000: 830), requiring ever more sophisticated computational tools and integrated knowledge systems (Zweiger 2001: 13).

Thirdly, the shift in the nature of 'knowledge' from molecular function to contextual function, with little clarity about what form this new knowledge will take, has also transformed the nature of 'boundary work' (Gieryn 1994) into a more collaborative rather than simply exclusionary exercise through the recognition that new knowledge cannot be achieved unless rigid demarcations between disciplines are significant eroded. 'An interdisciplinary spirit will come to guide those excited by the global analysis of protein function' (Fields 2001: 1224, Blackstock and Mann 2001). However, the changes are altogether much wider as such processes are embedded in broader societal institutions. For example, universities and research institutions in commerce and industry are finding the need to build multi-disciplinary life science centres as space within which to explore these new complexities, such as those in San Francisco's 'Biotech bay' (Zweiger 2001: 161) or the proposed 'genome fen' development at Hinxton outside Cambridge (Patel 1999).

Finally, translating the new genetics and molecular biology into a new paradigm requires more than just a technical understanding of gene structure and function, but also of 'epidemiology, biostatistics, environmental health science, health economics, management studies, medical informatics, information technology, and health services research' (Zimmern 1998:4). New technological options cannot survive in society without being entrenched in networks of producers, users and various services (Stemerding 1995: 144) including funding bodies, government and regulatory agencies, and firms as well as scientists. We are in effect witnessing the creation of a new research system, centred on the production, use and commodification of genetic knowledge, and based on new sets of knowledge, technologies and commodities, embodying a new set of sociotechnical relations involving new groups of actors (Martin 2001: 171). These as we saw in Chapters 5 and 8, include the wider public.

Members of the public, as citizens, require greater involvement in techno-scientific decision-making as we saw in Chapter 8, largely because science has become demystified, and its authority de-legitimised as part of the move to a 'risk' society (Beck 1992). Gibbons (1999: C82) argues that the practice of science is undergoing a shift from producing 'reliable' to producing 'socially robust' knowledge: 'science can no longer be validated as reliable by conventional discipline-bound norms; while remaining robust, science must now be sensitive to a much wider range of social implications'. Scientists must recognise that scientific work has already become transformed by the process of contextualisation with greater involvement by non-experts in techno-scientific decision-making as an integral part of producing socially robust knowledge (Nowotny, Scott and Gibbons 2001).

The preceding discussion suggests strongly that a paradigm shift in biotechnology is indeed occurring following the completion of the mapping and sequencing of the human genome. It suggests that future developments are likely to be located within new techno-scientific fields such as proteomics. The situated and co-constructed nature of techno-scientific work (see inter alia Bijker et al 1987, Bijker 1995) is well illustrated in the rapidly changing post-genomic era. The development of new technologies can thus no longer be conceived as simple, linear and deterministic (Ellul 1964). The relationships between the different elements of the situation (broadly conceived) within which post-genomic scientific work is accomplished are complex, multiple, dialectical, transformative, and even conflictual and contradictory (Clarke and Fujimura (1992: 6). How then should we look to the future with any confidence?

New Millennium, New Genetics?

In the UK, a White Paper called 'Our Inheritance, Our Future' was published in June 2003 by the Department of Health. It dealt with the government's view on the potential of genetics to improve healthcare within the ambit of the National Health Service (NHS). In a Foreword that underlined its importance, the Prime Minister, Tony Blair, suggested that the advances in genomic science would impact on society in this century as much as the computer did in the last. He then made a commitment to investing in 'research and research facilities to drive further discovery' and ensure that the NHS has the necessary skills and expertise to respond to the advances in genomics (DoH 2003: 1).

Building upon the completion of the Human Genome Project (HGP), the White Paper foresees a revolution in healthcare over the next few decades (DoH 2003: 7–8). More will be learnt about how individual genes or groups of genes interacting together can make us predisposed to certain diseases, and how some versions of genes can actually protect us from disease. We will also discover how our genes can affect our response to medicines, how genetic testing can become an integral part of healthcare, and how diseases such as cancer work at a cellular level. In order to take full advantage of the momentum in research in the new genetics, the White Paper sets out a detailed plan of action and investment for the next three years.

The NHS will lay the foundations for success by strengthening the existing hubs of expertise, boosting capacity, modernising laboratories, and expanding the workforce. It will support new initiatives in genetics-based care in key disease areas, in secondary and primary care, and through national screening programmes. It will invest in further education and training at all levels, and develop information systems and the evidence base. Finally, it will generate new knowledge and applications through investment in genetic research and development. Running parallel with these investments will be a programme to broaden and deepen open debate about the ethical and social issues, and a commitment to 'increasing public understanding of genetics and ensuring public confidence through a robust and proportionate system of regulation' (DoH 2003: 23).

A more expansive and far-reaching vision for the future direction of genomics research was outlined in *Nature* by Francis Collins and his colleagues at the US National Human Genome Research Institute in April 2003. Fifty years after the discovery of the double helical structure of DNA, identification of the genes responsible for human mendelian diseases can now be accomplished 'by a single graduate student with access to DNA samples and associated phenotypes, an Internet connection to the public genome databases, a thermal cycler and a DNA-sequencing machine' (Collins et al 2003: 835). Hence, as they note, their vision to capitalise on the HGP's immense potential to improve human health and well being addresses a different world than existed at the start of the mapping of the human genome. However, establishing 'robust paths from genomic information to improved human health', they admit, remains an immense challenge. Collins et al develop a blueprint for meeting this challenge based on three themes, genomics to biology, genomics to health and genomics to society, which they see as three floors of a building that rests firmly on the foundation of the HGP.

They also suggest six cross-cutting elements that run vertically through each of the thematic 'floors' rather like separate wings of a building. These include resources (the toolkits for genomic research), technology development (such as using nanotechnology), computational biology (including new ontologies and sophisticated management systems), training (in computational skills and interdisciplinarity), ethical, legal and social implications (through improvement of public accessibility and oversight), and education (making everyone more knowledgeable, from the public to health professionals).

Within each of the main crosscutting themes, Collins and his colleagues develop a number of grand challenges for the future, which effectively form a research agenda for the new genetics in this new millennium. In the genomics to biology theme, they centre on elucidating the structure and function of genomes. In the genomics to health theme, they are about translating genome-based knowledge into health benefits, and in the genomics to society theme, they are concerned with promoting the use of genomics to maximise benefits and minimise harms. Their vision concludes with some 'quantum leaps' (Collins et al 2003: 846) that speculate on the potential technical developments that might enhance research and clinical application so as to rewrite entire approaches to biomedicine. However, they also warn that progress will be impeded if there is no immediate release of publicly accessible data from large-scale sequencing projects, a lesson learned, as we have already noted in Chapter 3, from the HGP.

Recent qualitative social science research (much of which we have already included in this book) leads us to conclude that neither the British nor the American visions of the future for genomics research are likely to develop in quite the straightforward manner envisaged. For example, Webster (2003), in his consideration of the White Paper, refers to several areas that are particularly relevant. The long-discounted linear model of research and development, rooted as it is in the modernist strategy, still seems to underpin both visions. Little understanding is in evidence of the complexities surrounding such issues as the standardisation of measures, or the reality of informed consent in the context of understanding risk. The difficulties in balancing national instruments and measures, the quality of life assumptions embedded in them, and their local translations, have been largely overlooked. Finally, there is the issue of resources, and its link to public perceptions and anxieties in trying to second-guess future areas of high demand for genetic services.

Together, these reservations suggest that there is still an important role for social science research in informing the development of the policy process associated with this important area. Technosciences do not have their own built-in momentum allowing them to pass untouched through a neutral social medium (Bijker and Law 1992). Instead technological change is messily contingent, and born out of conflict, difference and resistance. Its products, for example a genetic test, should be seen as 'hybrids of human actors, natural phenomena and socio-technical systems', and are never simply just 'human, natural or socio-technical' (Webster 2003: 2).

Bibliography

Abbott A (1992) 'New French genome centre aims to prove that bigger is better' *Nature* 357, 226–227.

Adam B (1998) *Timescapes of Modernity* London, Routledge.

Adam B (2000) 'The temporal gaze: the challenge for social theory in the context of GM food' *British Journal of Sociology* 51, 1 (January/March), 125–142.

AEBC (2001) *Crops on Trial* London, Agriculture and Environment Biotechnology Commission.

AEBC (2002) *A Debate about the Issue of Possible Commercialisation of GM Crops in the UK* London, Agriculture and Environment Biotechnology Commission.

Affymetrix (2003) www.affymetrix.com.

Anderson C (1993) 'Genome shortcut leads to problems' *Science* 25: 1684–1687.

Andrews L and Nelkin D (2001) *Body Bazaar: the Market for Human Tissue in the Biotechnology Age* New York, Crown.

Anon (2002) 'Reaching their goal early sequencing labs celebrate' *Science*, 300, 409.

Anon (2001) 'Differentiation and integration' *Nature Biotechnology*, 19, July 2001, 597.

Atkinson D (1999) 'Glaxo awaits Relenza verdict' *The Guardian* Monday October 4, 20.

Balmer B (1993) *Mutations in the Research System. The Human Genome Project and Science Policy* D Phil thesis, Falmer, University of Sussex.

Balmer B (1996) 'The Political Cartography of the Human Genome Project' *Perspectives on Science* 4, 3, 249–282.

Barnes B and Edge D (Eds.) (1982) *Science in Context. Readings in the Sociology of Science* Milton Keynes, Open University Press.

Barns I (1996) 'Manufacturing Consensus? Reflections on the UK National Consensus Conference on Plant Biotechnology' *Science as Culture* 23, 200–217.

Bauer MW and Gaskell G (2002) *Biotechnology. The Making of a Global Controversy* Cambridge, Cambridge University Press.

Bauman Z (1993) *Postmodern Ethics* Oxford, Blackwell.

Bauman Z (1998) *Globalization. The Human Consequences* Cambridge, Polity.

Beck U (1992) *Risk Society: Towards a New Modernity* London, Sage.

Beck U (1995) *Ecological Politics in an Age of Risk* Cambridge, Polity.

Beck U (1998) 'Politics of Risk Society' in *The Politics of Risk Society*, Franklin, J (Ed.), Cambridge, Polity.

Beck U, Giddens A and Lash S (1994) *Reflexive Modernisation* Cambridge, Polity.

Bennett D, Glasner P, and Travis D (1986) *The Politics of Uncertainty: Regulating Recombinant DNA Research in Britain* London, Routledge and Kegan Paul.

Bernal JD (1939) *The Social Function of Science* London, Routledge.

Bijker WE (1995) 'Sociohistorical Technology Studies' in S Jasanoff et al (Eds.) *Handbook of Science and Technology Studies* Thousand Oaks, Sage, 229–256.

Bijker WE, Hughes TP and Pinch TJ (Eds.) (1987) *The Social Construction of Technological Systems: New Directions in the Sociology and History of Technology* Cambridge Massachusetts, MIT Press.

Bijker WE And Law J (Eds.) (1992) *Shaping Technology/ Building Society, Studies in Sociotechnical Change* Cambridge, Massachusetts, MIT Press.

Biospace (2000) *Genomics Primer*, www.biospace.com/articles/genomic.primer.print.cfm.

Bishop JE and Waldholz M (1990) *Genome* New York, Simon and Schuster.

Blackstock W and Mann M (2001) 'A boundless future for proteomics' in *TRENDS in Biotechnology* 19, 10 (Supplement), S1–2.

Bloomfield BP and Verdubakis T (1995) 'Disrupted Boundaries: New Reproductive Technologies and the Language of Anxiety and Expectation' *Social Studies of Science* 25, (3); 533–551.

BMA (1999) *The Impact of Genetic Modification on Agriculture, Food and Health – an Interim Statement* London, British Medical Association, May. [http://www.bma.org.uk/public/science/genmod.htm].

Bodmer JW (1993) 'The telephone book of life' *Nature*, 361, 580.

Bowker GC and Star SL (1999) *Sorting Things Out. Classification and its Consequences* Cambridge Massachusetts, MIT Press.

Bradby H (1996) Genetics and Racism, in Marteau TM and Richards MPM (Eds.) *The Troubled Helix: Social and Psychological Implications of the New Human Genetics* Cambridge, Cambridge University Press.

Bradford TC (2003) 'Evolving symbioses – venture capital and biotechnology' *Nature Biotechnology* 21, September, 983–984.

Brown P (2000) 'Lawyer's challenge to US over GM safety claims' *The Guardian* Tuesday February 29, 5.

Bud R (1993) *The Uses of Life: a History of Biotechnology* Cambridge, Cambridge University Press.

Burris J, Cook-Deegan R and Alberts B (1988) 'The Human Genome Project after a decade: policy issues' *Nature Genetics* 20, 20 December, 333–335.

Butler D (2002) 'Piecing it all together' *Nature* 420, 460.

Callon M (1986) 'Some elements in the sociology of translation: Domestication of the scallops and the fishermen of St Brieux Bay' in *Power, Action and Belief: a New Sociology of Knowledge?* Law, J (Ed.), London, Routledge and Kegan Paul.

Cambrosio A and Limoges C (1991) 'Controversies as Governing Processes in Technology Assessment' *Technology Analysis and Strategic Management* 3 (4), 377–396.

Campbell D (1999) 'US halts gene tests after youth dies' *The Guardian* Thursday 30 September, 19.

Carter S (1995) 'Boundaries of Danger and Uncertainty: an analysis of the technological culture of risk assessment' in Gabe L (Ed) *Medicine, Health and Risk: Sociological Approaches* Oxford, Blackwell.

Castells M (1997) *The Power of Identity* Oxford, Blackwell.

Celera (2003), www.celera.com.

Chitty M (2003) www.genomicglossaries.com.

Clarke AE and Fujimura JH (1992) 'What Tools? Which Jobs? Why Right?' in Clarke AE and Fujimura JH (Eds.) *The Right Tools for the Job. At Work in Twentieth-Century Life Sciences* Princeton, Princeton University Press.

Close F (1992) *Too Hot to Handle: the Race for Cold Fusion* London, WH Allen.

Cochrane A (1996) 'From Theories to Practices: Looking for Local Democracy in Britain' in *Rethinking Local Democracy* King, D and Stoker, G (Eds.), Basingstoke, Macmillan.

Cohen D (1993) *Les Genes de l'Espoir: A la Decouverte du Genome Humain* Paris, Robert Laffont.

Collier DA (2001) 'Sudden impact? The human genome sequence and the pace of gene discovery in complex diseases' *The Pharmacogenomics Journal* 1, 9–11.

Collins FS and Gallas D (1993) 'A new five year plan for the U.S. Human Genome Project' *Science* 262.

Collins FS, Morgan M and Patrinos A (2003) 'The Human Genome Project: lessons from large-scale biology' *Science* 300, 11 April, 286–290.

Collins F et al (2003) 'A vision for the future of genomics research' *Nature* 422, 24 April, 835–847.

Collins HM (1985) *Changing Order. Replication and Induction in Scientific Practice* London, Sage.

Collins HM (1988) 'Public experiments and displays of virtuosity: the core set revisited' *Social Studies of Science* 18, 725–748.

Collins HM (Ed.) (1981) 'Special Issue Knowledge and Controversy: Studies of Modern Natural Science' *Social Studies of Science* 11, 1–158.

Collins HM and Pinch T (1982) *Frames of Meaning. The Social Construction of Extraordinary Science* London, Routledge and Kegan Paul.
Conner S (1999) 'Pusztai: the verdict' *The Independent* Friday Review, 19 February, 9.
Cook-Deegan R, Chan C and Johnson A (2000) *World Survey of Funding for Genomics Research.* Final Report to the Global Forum for Health Research and the World Health Organisation. September, Washington D.C.
Cook-Deegan RM (1989) 'The Alta Summit, December 1984' *Genomics* 5, 661–663.
Cook-Degan R (1994) *The Gene Wars: Science, Politics and the Human Genome* New York, Norton.
Cooper NG (Ed) (1994) *The Human Genome Project: Deciphering the Blueprint of Heredity* Mill Valley CA, University Science Books.
Coote A and Lenaghan J (1997) *Citizens' Juries: Theory into Practice* London, Institute for Public Policy Research.
Courteau J (1991) 'Genome Databases' *Science* 254: 201–207.
Cousens, SN et al (1997) 'Predicting the CJD epidemic in humans' *Nature* 385, 16 January, 197–198.
Crick F (1990) *What Mad Pursuit* Harmondsworth, Penguin Books.
Crook S, Pakulski J and Waters M (1992) *Postmodernization. Change in Advanced Society* London, Sage.
Curagen (2003), www.curagen.com.
Davies K (2001) *The Sequence: Inside the Race for the Human Genome* London, Weidenfeld and Nicolson.
Davis J (1990) *Mapping the Code: the Human Genome Project and the Choices of Modern Science* New York , John Wiley.
Dawes B (1952) *A Hundred Years of Biology* London, Duckworth.
De Waele D (1997) 'The virtual reality of the biotechnology debate' in S Sterckx (Ed) *Biotechnology, Patents and Morality* Aldershot, Ashgate.
DeLisi C (1988) 'The Human Genome Project', *American Scientist* 76, 488–493.
Dodson M and Rothwell R (1994) *The Handbook of Innovation* Aldershot, Edward Elgar.
DoH (2003) *Our Inheritance, Our Future. Realising the potential of genetics in the NHS* London, Department of Health.
Dooley JF (2002) *The Genomics Outlook to 2005: Transforming Pharmaceutical and Diagnostic Markets* London, Reuters Business Insight.
Douglas M (1995) *Risk Acceptability According to the Social Sciences* London, Routledge and Kegan Paul.
Driscoll M and Carr-Brown J (1999) 'What's eating us?' *The Sunday Times, News Review* 21 February, 6.
DTI (1998) *The Implications of Genetic Testing for Insurance* London, Department of Trade and Industry, Department of Health, and Office of Science and Technology.
Dulbecco R (1986) 'A turning point in cancer research' *Science* 231, 1055–1056.
Dunham I et al (1999) 'The DNA sequence of human chromosome 22' *Nature* 402, 489–495.
Dunkerley D and Glasner P (1998) 'Empowering the Public? Citizens' Juries and the New Genetic Technologies' *Critical Public Health* 8, 181–192.
Dunn LC (1991) *A Short History of Genetics* Ames, Iowa, Iowa State University Press.
Eden S (1996) 'Public participation in environmental policy: considering scientific, counter-scientific and non-scientific contributions' *Public Understanding of Science* 5, 183–204.
Edmond G and Mercer D (1997) Scientific literacy and the jury: reconsidering jury competence *Public Understanding of Science* 6, 329–357.
Evans G and Durant J (1989) 'Understanding of Science in Britain and America' in Howell R et al (Eds.) *British Social Attitudes: Special International Report* Aldershot, Gower.
Eisenberg D., Marcotte EM, Xenarious I and Yeates TO (2000) 'Protein Function in the post-genomic era' *Nature* 405, 15 June, 823–826.
Elliott L (2003) 'The lost decade', in *The Guardian* Wednesday July 9, 1–2.
Ellul J (1964) *The Technological Society* New York, Alfred Knopf.
Elzinga A and Jamison A (1995) 'Changing Policy Agendas in Science and Technology' in Jasanoff S et al (Eds.) *Handbook of Science and Technology Studies* Thousand Oaks, Sage.

Elzinga A (2002) 'The new production of reductionism in models relating to research policy' *Science and Industry in the 20th Century* Stockholm, Nobel Symposium, November, 21-23.
EN (2002) 'MJ sues over DNA sequencer' *Nature Biotechnology* 20 July, 647.
Englehardt TH and Caplan AL (Eds.) (1987) *Scientific Controversies: Case Studies in the Resolution and Closure of Disputes in Science and Technology* Cambridge, Cambridge University Press.
Ernst and Young (2003) *Beyond Borders. The Global Biotechnology Report 2003* London, Ernst and Young International.
Etzkowitz H (2002) 'The triple helix: the entrepreneurial university and the industrialisation of research', Abstract NS 123–5, *Science and Industry in the 20th Century* Stockholm, Nobel Symposium, November, 21–23.
Etzkowitz H and Webster A (1995) 'Science as intellectual property' in Jasanoff S et al (Eds.) *Handbook of Science and Technology Studies'* Thousand Oaks, Sage, 480–505.
Etkowitz H, Webster A, Healey P (Eds.) (1998) *Capitalising Knowledge: New Intersections of Industry and Academia* Albany, State University of New York Press.
Evans GA (1993) 'MegaYAC library' *Science* 260: 677.
Fairhall D (1997) 'Ministers misled over Gulf' *The Guardian* February 27, 6.
Fields C (1992) 'Data exchange and inter-database communication in genome projects' *Tibtech* 10: 58–61.
Fields S (2001) 'Proteomics in genomeland' *Science* 219, 16 February, 1221–1224.
Fisken J, Rutherford J (2002) 'Business models and investment trends in the biotechnology industry in Europe' *Journal of Commercial Biotechnology* 8, (3), 191–199.
Fixdal J (1997) 'Consensus Conferences as "Extended Peer Groups"' *Science and Public Policy* 24, (6); 366–376.
Fox S (1991) 'Applications for synthesising and sequencing DNA beyond the Genome Project' *Genetic Engineering News*, June, 6–8.
Frazier ME, Johnson GM, Thomassen DG, Oliver CE, Patrinos A (2003) 'Realising the potential of the genome revolution: the genomes to life programme' *Science* 300, 11 April, 290–293.
Frederickson RM (2003) 'Biobusiness on the web' *Nature Biotechnology,* 21, May, 499–503.
Fujimura JH (1987) 'Constructing "do-able" problems in cancer research: articulating alignment' *Social Studies of Science* 17, 257–293.
Fujimura J (1999) 'The Practices of Producing Meaning in Bioinformatics' in Fortun M and Mendelsohn E (Eds.) *The Practice of Human Genetics* Sociology of the Sciences Yearbook, 1997, Dordrecht, Kluwer.
Fukuyama F (2002) *Our Posthuman Future: Consequences of the Biotechnology Revolution,* London, Profile Books.
Funtowicz SO and Ravetz, J (1993) 'Science for the post-normal age' *Futures* 25 (7), 739–755.
Gaskell G et al (1999) 'Worlds Apart? The Reception of Genetically Modified Foods in Europe and the US' *Science* 285, 16 July, 384–387.
Gibbons M (1999) 'Science's new social contract with society' *Nature* 402, supplement 2 December, C81–84.
Gibbons M, Limoges C, Nowotny H, Schwartzman S, Scott P and Trow M (1994) *The New Production of Knowledge. The Dynamics of Science and Research in Contemporary Societies* London, Sage.
Giddens A (1989) *Sociology* Cambridge, Polity.
Giddens A (1990) *The Consequences of Modernity* Cambridge, Polity.
Giddens A (1991) *Modernity and Self-identity: Self and Society in the Late Modern Age* Cambridge, Polity.
Giddens A (1998) *The Third Way* Cambridge, Polity.
Giddens A (1998) 'Risk Society: the Context of British Politics' in *The Politics of Risk Society* Franklin, J. (Ed.), Cambridge, Polity.
Gieryn TF (1983) 'Boundary work and the demarcation of science from non-science: Strains and interests in professional ideologies of scientists' *American Sociological Review* 48, 781–795.

Gieryn TF (1995) 'Boundaries of Science' in Jasanoff S et al (Eds.) *Handbook of Science and Technology Studies* Thousand Oaks, Sage, 394–443.

Gilbert GN and Mulkay M (1984) *Opening Pandora's Box. A Sociological Analysis of Scientists' Discourse* Cambridge, Cambridge University Press.

Glasner P (1996) 'From community to "collaboratory"? The Human Genome Mapping Project and the changing culture of science' *Science and Public Policy* 2, April: 109–116.

Glasner P (2002) 'Beyond the genome: reconstituting the new genetics' in *New Genetics and Society* 21, 3, 267–277.

Glasner P and Dunkerley D (1999) 'The new genetics, public involvement, and citizens' juries: a Welsh case study' *Health, Risk and Society* 1, 3, 313–324.

Glasner P and Rothman H (1999) 'Does Familiarity Breed Concern? Bench Scientists and the Human Genome Mapping Project' *Science and Public Policy* 26, 313–324.

Glasner P and Rothman H (2001) 'New genetics, new ethics? Globalisation and its discontents' *Health, Risk and Society* 3, 3, 245–259.

Glasner P and Rothman H (2004) 'What's so new about the "new genetics"?' in P Glasner (Ed.) *Reconfiguring Nature. Issues and Debates in the New Genetics* Aldershot, Ashgate.

Glasner P, Rothman H and Travis D (1995) 'Exploring organisational issues in British genomic research' *The Genetic Engineer and Biotechnologist* 15, 2/3, 124–133.

Goldberg R (1999) 'The business of agriceuticals' *Nature Biotechnology* 17, March, Supplement BV5–6.

Goozner M (2000) 'Patenting life' *The American Prospect*, 11 (26).

Green P (1997) 'Against a whole-genome shotgun' *Genome Research,* 7, 410–417.

Halbeck M (2003) 'Brussels takes EU states to court over biopatent law' *Nature Biotechnology* 21, Sept., 960.

Hall SS (1988) *Invisible Frontiers: the Race to Synthesise a Human Gene* London, Sidgwick and Jackson.

Hamilton N (1998) *Attack of the Genetically Engineered Tomatoes. The Ethical Dilemma of the '90s* Stowmarket, Whittet Books/Nemesis Press.

Hampel J and Renn O (2000) 'Introduction: public understanding of genetic engineering, in German Attitudes to Genetic Engineering' Special Issue of *New Genetics and Society* 19, 2, 221–231.

Haraway DJ *Simians, Cyborgs and Women: The Reinvention of Nature* London, Routledge.

Harrison S and Mort M (1998) 'Which Champions, Which People? Public and User Involvement in Health Care as a Technology of Legitimation' *Social Policy and Administration* 32, 1, 60–70.

HGAC (1999) *The Implications of Genetic Testing for Employment* London, The Human Genetics Advisory Commission.

HGC (2000) *Whose Hand on Your Genes* London, Human Genetics Commission.

HGC (2002) *Inside Information. Balancing Interests in the Use of Personal Genetic Data* London, Human Genetics Commission.

Hilgartner S (1995) 'The Human Genome Project', in Jasanoff S. et al (Eds.) *Handbook of Science and Technology Studies* Thousand Oaks, Sage, 302–315.

Hill A and Michael M (1998) 'Engineering Acceptance: Representations of "The Public" in Debates on Biotechnology' in Wheale P, von Schomberg R and Glasner P (Eds.) *The Social Management of Genetic Engineering* Aldershot, Ashgate.

Hine C (1998) 'Information technology as an instrument of genetics', in P Glasner and H Rothman (Eds.) *Genetic Imaginations. Ethical, Legal and Social Issues in Human Genome Research* Aldershot, Ashgate.

Hirst P and Thompson G (1996) *Globalization in Question: the International Economy and the Possibilities of Governance* Cambridge, Polity.

Hood L and Galas D (2003) 'The digital code of DNA' *Nature* 421, 23 January, 444–448.

Horizon (1998) *The Book of Man* London, BBC.

House of Lords (2000) *Science and Society* HL Paper 38, London, House of Lords Select Committee on Science and Technology.

Housman D and Ledley FD (1998) 'Why Pharmacogenomics? Why now?' *Nature Biotechnology* 16, June, 492–493.

Huang S (2000) 'The practical problems of post-genomic biology' *Nature Biotechnology* 18, May, 471–472.
Hubbard R (1986) 'Eugenics and prenatal testing', *International Journal of Health Services* 16, 2, 227–242.
Human Genome News (1999) 10, 1–2, February.
Human Genome News (1996) 7(6), April-June.
Human Genome Sciences (2003), www.hgsi.com.
Incyte (2003), www.incyte.com.
International Human Genome Sequencing Consortium (2001) 'Initial sequencing and analysis of the human genome' *Nature* 409, 15 February, 860–921.
Iredale R and Longley M (1997) *Citizens' Juries: Getting Non-Scientists Involved* mimeo, Glamorgan, Welsh Institute for Health and Social Care.
Irwin A (1995) *Citizen Science. A Study of People, Expertise and Sustainable Development* London, Routledge.
Irwin A and Wynne B (Eds.) (1996) *Misunderstanding Science? The Public Reconstruction of Science and Technology* Cambridge, Cambridge University Press.
Jasanoff S et al (Eds.) (1995) *Handbook of Science and Technology Studies* Thousand Oaks, Sage.
Jasanoff S (1995) 'Product, process or programme: three cultures and the regulation of biotechnology' in *Resistance to New Technology:Nuclear Power Information Technology and Biotechnology* Bauer, M (Ed.), Cambridge, Cambridge University Press.
Joss S and Brownlea A (1999) 'Considering the concept of procedural justice for public policy – and decision-making in science and technology' *Science and Public Policy* 26, 5, 321–330.
Joss S and Durant J (Eds.) (1995) *Public Participation in Science. The Role of Consensus Conferences in Europe* London, The Science Museum.
Kanehisa M and Bork P (2003) 'Bioinformatics in the post-sequence era' *Nature Genetics Supplement* 33 March: 305–310.
Kate KT and Laird SLA (1999) *The Commercial Use of Biodiversity: Access to Genetic Resources and Benefit Sharing* London, Earthscan.
Keating P, Limoges C and Cambrosio A (1999) 'The Automated Laboratory. The Generation and Replication of Work in Molecular Genetics', in Fortun M and Mendelsohn E (Eds.) *The Practices of Human Genetics* (Sociology of the Sciences Yearbook 1997) Dordrecht, Kluwer.
Kerr A, Cunningham-Burley S and Amos A (1997) 'The new genetics: professionals' discursive boundaries' *Public Understanding of Science* 4, 243–253.
Kerr A, Cunningham-Burley S and Amos A (1998) 'The New Genetics and Health: Mobilising Lay Expertise' *Public Understanding of Science* 7, 41–60.
Kleiner K (1999) 'Monarchs under siege' *New Scientist* 22 May, 4.
Krimsky S (1982) *Genetic Alchemy: the Social History of the Recombinant DNA Controversy* Cambridge, Massachusetts, MIT Press.
Lander E (1994) 'DNA Fingerprinting: Science, Law and the Ultimate Identifier' in DJ Kevles and L Hood (Eds.) *The Code of Codes. Scientific and Social Issues in the Human Genome Project* Cambridge Massachusetts, Harvard University Press.
Lash S, Szerszynski B and Wynne B (1996) *Risk, Environment and Modernity: Towards a New Ecology* London, Sage.
Latour B (1987) *Science in Action. How to Follow Scientists and Engineers Through Society* Milton Keynes, Open University Press.
Latour B and Woolgar S (1979) *Laboratory Life. The Social Construction of Scientific Facts* London, Sage.
LCGIS (1998) *Citizen Foresight: A Tool to Enhance Democratic Policy-making. Vol.1: The Future of Food and Agriculture* London: London Centre for Governance, Innovation and Science, and The Genetics Forum.
Lederberg J and Uncapher K (1989) *Towards a National Collaboratory. Report of an Invitational Workshop at the Rockefeller University* New York, National Science Foundation.

Lee TF (1991) *The Human Genome Project: Cracking the Genetic Code of Life* New York, Plenum Press.

Levidow L (1999) 'Britain's biotechnology controversy: elusive science, contested expertise' *New Genetics and Society* 18, 47–64.

Lewenstein BV (1995) 'From Fax to Facts: Communication in the Cold Fusion Saga' *Social Studies of Science* 25, 403–36.

Lewontin RC (1991) *The Doctrine of DNA: Biology as Ideology* Harmondsworth, Penguin.

Lewontin R (2001) 'They got the wrong key of life' *The Sunday Times* (News Review) July 8, 4.

Lockhart DJ and Winzeler EA (2000) 'Genomics, gene expression and DNA arrays' *Nature* 405, 15 June, 827–836.

Lynch M and Jasanoff S (Eds.) (1998) 'Contested Identities: Science Law and Forensic Practice' *Social Studies of Science* (Special Issue) 28, 5–6.

Macnaghten P and Urry J (1998) *Contested Natures* London, Sage

Malakoff D (2003) 'Academia gets no help from U.S. in patent case' *Science* 300, 13 June, 1635–1636.

Marlier E (1992) 'Eurobarometer 35.1: opinions of Europeans on biotechnology in 1991' in Durant J *Biotechnology in Public: a Review of Recent Research* London, The Science Museum.

Marshall E (2000) 'Clinton and Blair back rapid release of data' *Science* 287, 17 March: 1903.

Marshall E (1999) 'Drug firms to create public data-base of genetic mutations' *Science* 284, 16 April, 406–407.

Marshall E (2000) 'Genome sequencing: talks of a public-private deal end in acrimony' *Science* 287, 10 March, 1723–1725.

Marshall E (2001) 'Bermuda rules: community spirit, with teeth' *Science* 291, 16 Feb, 192.

Marshall E and Pennisi E (1988) 'Hubris and the human genome' *Science* 280, 15 May, 994–995.

Martin P (1995) 'The American Gene Therapy Industry and the Social Shaping of a New Technology' *The Genetic Engineer and Biotechnologist* 15 (2/3), 155–167.

Martin P (2001) 'Genetic governance: the risks, oversight and governance of genetic databases in the UK' *New Genetics and Society* 20, 2, 157–183.

Martin P and Kaye J (2000) 'The use of large biological sample collections in genetics research: issues for public policy' *New Genetics and Society* 19, 2, 165–191.

Marx K (1973) *Grundrisse* London, Allen Lane.

Marx K (1976) *Capital Vol 1*, Harmondsworth, Penguin.

Mazur A (1981) *The Dynamics of Technical Controversy* Washington DC, Communications Press.

McCain KW (1991) 'Communication, competition and secrecy: the production and dissemination of research-related information in genetics' *Science, Technology and Human Values* 16 (4): 491–516.

McIver S (1997) *An Evaluation of the King's Fund Citizens' Juries Programme* Birmingham, Health Services Management Centre, University of Birmingham.

McMillan GS, Narin F, Deeds DL (2000) 'An analysis of the critical role of public science in innovation: the case of biotechnology' *Research Policy* 29, 1, 1–8.

McNally R (1998) 'Eugenics Here and Now' in Glasner P and Rothman H (Eds.) *Genetic Imaginations. Ethical, Legal and Social Issues in Human Genome Research* Aldershot, Ashgate.

Meek J. '$500m human map to trump the DNA project' London, *The Guardian*, 6 April.

Merton RK (1973) *The Sociology of Science. Theoretical and Empirical Investigations* Chicago, University of Chicago Press.

Meyers EW et al (2000) 'A whole-genome assembly of Drosophila' *Science*, 24 March, 2196–2204.

Michael M (1996) *Constructing Identities* London, Sage.

Millennium Pharmaceuticals (2003), www.minm.com.

Mitchell P (2002) 'European Commission rethinks biotech patents' *Nature Biotechnology*, December, 1175–1176.

Mulkay M (1979) *Science and the Sociology of Knowledge* London: George Allen and Unwin.
National Research Council, Committee on Mapping and Sequencing the Human Genome. (1988) *Mapping and Sequencing the Human Genome* Washington D.C., National Academy Press.
Nature (2000) 'Greens pursuade Europe to revoke patent on neem tree' *Nature* 405 18 May, 266–267.
Nature (2001) International Human Genome Sequencing Consortium 'Initial sequencing of the human genome' *Nature* 409 (15 February), 860–921.
Nature Biotechnology (2000) 'Intellectual Propriety' *Nature Biotechnology* 18 (5) 469.
Nature Insight (2000) 'Insight: functional genomics' *Nature* 405 15 June, 819–865.
Nelkin D (1975) 'The Political Impact of Technical Expertise' *Social Studies of Science* 5, 35–44.
Nelkin D (1982) 'Controversy as a political challenge' in Barnes B and Edge D (Eds.) *Science in Context: Readings in the Sociology of Science* Milton Keynes, Open University Press.
Nelkin D (1995) 'Science Controversies: the Dynamics of Public Disputes in the United States' in *Handbook of Science and Technology Studies* Jasanoff, S et al (Eds.), London, Sage.
Nelkin D and Andrews L (1999) 'DNA identification and surveillance creep' in P Conrad and J Gabe (Eds.) *Sociological Perspectives on the New Genetics* Oxford, Blackwell.
Nelkin D and Lindee S (1995) *The DNA Mystique: The Gene as a Cultural Icon* New York, Freeman.
Nowotny H, Scott P and Gibbons M (2001) *Rethinking Science. Knowledge and the Public in an Age of Uncertainty* Cambridge, Polity.
Nuffield Council on Bioethics (1999) *Genetically Modified Crops: the Ethical and Social Issues* London, Nuffield Council on Bioethics.
Nuffield Council on Bioethics (2002) *The Ethics of Patenting DNA* London, Nuffield Council on Bioethics.
Nuki P (1999) 'GM food advisers have links to biotech companies' *The Sunday Times* 13 June, 5.
OECD (1999) *Agricultural Policies in OECD Countries. Monitoring and Evaluation* Paris, OECD.
Office of Technology Assessment (1988) *Mapping Our Genes* Washington D.C., Government Printing Office.
Oliver M (1996) *Understanding Disability: From Theory to Practice* Basingstoke, Macmillan.
Olson M, Hood L, Cantor C and Botstein D (1989) 'A common language for physical mapping of the human genome' *Science* 245, 1434–1435.
Orsenigo L (1989) *The Emergence of Biotechnology* London, Pinter.
Paabo S (2001) 'The human genome and our view of ourselves' *Science* 291, 16 February, 1219–1220.
Papadopolos S (2000) *Nature Biotechnology* 18 Suppl., IT3–IT4.
Parsons E and Atkinson P (1992) 'Lay constructions about genetic risk' *Sociology of Health and Illness* 14, 437–455.
Patel K (1999) 'Whitehall split over Hinxton bid' *The Times Higher Education Supplement*, 17 September, 7.
Pateman C (1970) *Participation and Democratic Theory* Cambridge, Cambridge University Press.
PEALS (2003) *Verdict is out from first ever 'DIY Citizens' Jury'*, Press Release, 15 January, Newcastle, University of Newcastle Institute for Policy, Ethics and the Life Sciences.
Pearson H (2003) 'Geneticists play the numbers game in vain' *Nature* 423, 5 June, 576.
Peltonen L (2001) 'The molecular dissertation of human diseases after the human genome' *The Pharmacogenomics Journal* 1, 5–14.
Peltonen L and McKusick V A (2001) 'Dissecting human disease in the postgenomic era' *Science* 291, 16 February, 1224–1229.

Pennisi E (1999) 'Human Genome: academic sequencers challenge Celera in a sprint to the finish' *Science*, 283, 16 April, 406–407.

Pennisi E (2000) 'Fruit fly genome yields data and a validation' *Science* 287, 25 February, 374.

Perelman M (2002) *Steal This Idea: Intellectual Property Rights and the Corporate Confiscation of Creativity* New York, Palgrave.

Persidis A (1998) 'Bioentrepreneurship around the world' *Nature Biotechnology* 16, May, Supplement, 3–4.

Persidis A (1999) 'Agricultural biotechnology' *Nature Biotechnology* 17, 6, 612–614.

Phillips A (1994) 'Plurality, Solidarity and Change' in Weeks J (Ed) *The Lesser Evil and the Greater Good* London, Rivers Oram.

Pickard S (1998) 'Citizenship and Consumerism in Health Care: A Critique of Citizens' Juries', *Social Policy and Administration* 32, 3, 226–244.

Pilnick A (2002) *Genetics and Society. An Introduction* Buckingham, Open University Press.

Pinch T and Bijker WE (1987) 'The Social Construction of Facts and Artefacts: Or How the Sociology of Science and the Sociology of Technology Might Benefit Each Other', in Bijker WE, Hughes TP and Pinch TJ (Eds.) *The Social Construction of Technological Systems: New Directions in the Sociology and History of Technology* Cambridge, Massachusetts, MIT Press.

POST (2001) *Open Channels. Public Dialogue in Science and Technology* Report 153, London, Parliamentary Office of Science and Technology.

Poste G (1998) *Nature Biotechnology* 16 (Supplement).

Price D de Solla (1963) *Little Science, Big Science* New York, Columbia University Press.

Price DJ de Solla (1984) 'The science/technology relationship, the craft of experimental science, and policy for the improvement of high technology invention' *Research Policy* 13, 3–20.

Purdue D (1995) 'Whose Knowledge Counts? "Experts", "Counter-experts" and the "Lay Public"' *The Ecologist* 25, (5), 170–172.

Purdue D (1999) 'Experiments in the governance of science and technology: a case study of the UK National Consensus Conference' *New Genetics and Society* 18, 79–99.

Putney SD, Herlihy WC, Schimmel P (1983) 'A new troponin T and cDNA clones for 13 different muscle proteins formed by shotgun sequencing' *Nature* 302, 713.

Radford T (1999) 'They don't know, you know' *The Guardian* 23 February, 17.

Rasnick D (2003) 'The biotechnology bubble machine' *Nature Biotechnology* 21, April, 355–356.

Ratchford JT and Columbo U (1996) 'Megascience' *UNESCO World Science Report*, 214–222.

Ravetz JR (1971) *Scientific Knowledge and its Problems* Harmondsworth, Penguin.

Richards M (1996) 'Family, Kinship and Genetics' in Marteau T and Richards M (Eds.) *The Troubled Helix. Social and Psychological Implications of the New Genetics* Cambridge, Cambridge University Press.

Rifkin J (1998) *The Biotech Century. How Genetic Commerce Will Change the World* London, Phoenix.

Rip A (1986) 'Controversies as informal technology assessment' *Knowledge* 8 (December), 349–371.

Rodinson M (1977) *Islam and Capitalism* Harmondsworth, Penguin.

Roos DS (2001) 'Bioinformatics – trying to swim in a sea of data' *Science* 291, 16 February, 1260–1261.

Rose H (1994) *Love, Power and Knowledge,* Cambridge, Polity Press.

Rose H (2001) 'Gendered genetics in Iceland' *New Genetics and Society* 20, 2, 119–138.

Rose N (1996) *Inventing Our Selves* Cambridge, Cambridge University Press.

Rosenberg A (1998) 'The human genome project: research tactics and economic strategies' in DL Hull and M Ruse (Eds.) *The Philosophy of Biology* Oxford, Oxford U.P., 567–585.

Ross JM and Ross M (1926) 'Grail, Holy' in Hastings J (Ed) *Encyclopaedia of Religion and Ethics* Edinburgh, T and T Clark, 385–389

Rothman BK (1986) *The Tentative Pregnancy: prenatal diagnosis and the future of motherhood* London, Allen and Unwin.

Rothman BK (1995) 'Of maps and imaginations: sociology confronts the genome' *Social Problems* 42, 1, 1–10.
Rothman H (1994) 'Between science and industry: the Human Genome project and instrumentalities' *The Genetic Engineer and Biotechnologist* 14, 2, 81–91.
Rothman H (1998) 'Gene sequencer' in Bud R. and Warner DJ (Eds.), *Instruments of Science: An Historical Encyclopaedia* New York, Garland, 281–283.
Rothman H, Glasner P and Adams C (1996) 'Proteins, plants and currents: rediscovering science in Britain' in *Misunderstanding science? The public reconstruction of science and technology*, Irwin A and Wynne B (Eds.) Cambridge, Cambridge University Press.
Rowen L, Mahairas G and Hood L (1997) 'Sequencing the human genome' *Science* 278, 24 Oct, 1605–1607.
Royal Society (1985) *The Public Understanding of Science* London, The Royal Society.
Royal Society (1999) *Review of data on possible toxicity of GM potatoes.* [http://www.royalsoc.ac.uk/st_pol54.htm].
Rubin G and Lewis EB (2000) 'A brief history of Drosphila's contributions to genome research' *Science* 287, 24 March, 2216–2218.
Sagar A, Daemmrich A and Ahiya M (2000) 'The tragedy of the commoners: biotechnology and its publics' *Nature Biotechnology* 18, 1, 2–4.
Sample I (203) 'The man in the street gets his forum on GM food – but decides to stay in the street' in *The Guardian*, Wednesday 4 June, 3.
Schatz BR (1991) 'Building an electronic community system' *Journal of Management Information Systems* 8, 87–107.
Schindler M (1992) 'Instruments and the progress of science' *Science* 258, 1423.
Science (2001) Venter C et. al., 'The sequence of the human genome' *Science* 291, 16 February, 1304–1351.
Scientific American Presents (1999) 'Your Bionic Future: How technology will change the way you live in the next millennium' *Scientific American Presents* 10, 3.
Scott A (Ed.) *The Limits of Globalization: Cases and Arguments* London, Routledge.
Saegusa A (1999) 'Japan pushes to capitalize on biotechnology' *Nature Biotechnology* 17, 4, 320–321.
Seargeant J and Steele J (Eds.) (1998) *Consulting the Public. Guidelines and Good Practice* London, Policy Studies Institute.
Shapin S (2003) 'Ivory trade' *London Review of Books*, 11 September, 15–19.
Shiva V (1991) 'Biodiversity, biotechnology and profits' in V Shiva et al (Eds.) *Biodiversity. Social and Ecological Perspectives* London, Zed Books.
Shorett P, Rabinow P and Billings PR (2003) 'The changing norms of the life sciences' *Nature Biotechnology* 21, February, 123–125.
Silver LM (1998) *Remaking Eden: Cloning and Beyond in a Brave New World* London, Weidenfeld and Nicolson.
Sinsheimer RL (1994) *The Strands of a Life: the Science of DNA and the Art of Education* Berkeley, University of California Press.
Smith R and Wynne B (Eds.) (1989) *Expert Evidence. Interpreting Science in the Law* London, Routledge.
Snedden R (2000) 'The Challenge of Pharmacogenetics and Pharmacogenomics' *New Genetics and Society* 19, 2, 145–164.
Snow CP (1981) *The Physicists: A Generation that Changed the World* London, Macmillan.
Soil Association (2002) *Seeds of Doubt. North American Farmers' Experiences of GM Crops* Bristol: Soil Association.
Spallone P and Wilkie T (2000) 'The research agenda in pharmacogenetics and biological sample collections – a view from the Wellcome Trust' *New Genetics and Society* 19, 2, 193–205.
SRO (1999) 'Rapid Response – the Genetic Modification of Food' *Sociological Research Online* 4, 3. [http://socresonline.org.uk]
Star SL and Griesemer JR (1989) 'Institutional Ecology, "Translations" and Boundary Objects: Amateurs and Professionals in Berkeley's Museum of Vertebrate Zoology, 1907–39' *Social Studies of Science* 19, 387–420.

Steinbrecher RA (2001) 'Ecological consequences of genetic engineering' in Tokar B (Ed) *Redesigning Life: The Worldwide Challenge to Genetic Engineering* London, Zed Books.
Steinmueller WE (1994) 'Basic research and industrial innovation' in Dodson M and Rothwell R (Eds.) *The Handbook of Innovation* Aldershot, Edward Elgar.
Stemerding D (1995) 'The Entrenchment of Human Genome Technology in Society: On Shifting Boundaries between Private and Public Discourses' in von Schomberg R (Ed.) *Contested Technology, Ethics, Risk, and Public Debate* Tilburg, International Centre for Human and Public Affairs.
Stewart J, Kendall E and Coote A (1994) *Citizens' Juries* London, IPPR.
Stock G (2002) *Redesigning Humans: Choosing our Children's Genes* London, Profile Books.
Strong P (1983) *The Ceremonial Order of the Clinic* London, Routledge and Kegan Paul.
Subcommittee on Energy and the Environment of the Committee on Science (2000) 'Written testimony of Gerald M. Rubin' U.S. House of Representatives, Washington DC 6 April.
Sulston J and Ferry G (2002) *The Common Thread: A Story of Science, Politics, Ethics and the Human Genome* London, Bantam Press.
Tate WT (2000) *Emergent Business Models* London, Council for Excellence in Management and Leadership.
Thomas SM, Hopkins MM and Brady M (2002) 'Shares in the human genome – the future of patenting DNA' *Nature Biotechnology* 20, December, 1185–1185.
Titmus RM (1970) *The Gift Relationship* London, Allen and Unwin.
United States Human Genome Project (1990) 'The First Five Years: Fiscal Years 1991–1995', see posted at www.genome.gov.
Vasil I (1998) 'Biotechnology and food security for the 21st century. A real-world perspective' *Nature Biotechnology* 16, May, 399–40.
Velody I and Williams R (Eds.) (1998) *The Politics of Constructionism* London, Sage.
Venter C 2000 'Statement of J. Craig Venter' *Subcommittee on Energy and Environment U.S. House of Representatives Committee on Science* 6 April, Washington D.C., House of Representatives.
Venter C et al (2001) 'The sequence of the human genome' *Science* 291, 16 February, 304–1351.
Venter C, Smith H and Hood L (1996) 'A new strategy for genome sequencing' *Nature* 381, 364–366.
Vines G (1997) 'Genetics: let the public decide' *British Medical Journal* 314, 5 April 1055.
Vucmirovic OG and Tighman SM (2000) 'Exploring genome space' *Nature* 405, 15 June, 820–822.
Wade N (1980) 'Court says lab-made life can be patented' *Science*, 208, 27 June, 1445.
Wakeford T (2002) 'Citizens' Juries: a radical alternative to social research' *Social Research Update* Issue 37, Guildford, University of Surrey.
Walsh G (2003) 'Biopharmaceutical benchmarks – 2003' *Nature Biotechnology* 21, August, 865–870.
Walsh V (2002) 'Biotechnology and the UK 2000–2005: globalization and innovation' in *New Genetics and Society* 21, 2, 149–176.
Waterston R and Sulston JE (1998) 'The Human Genome Project: reaching the finishing line' *Science* 283, 2 October, 53.
Weber JL and Myers EW (1997) 'Human whole-genome shotgun sequencing' *Genome Research* 7, 401–409.
Webster A (2003) 'The Government White Paper on Genetics: an Initial Comment' *Newsletter of the Innovative Health Technologies Programme* York, University of York Department of Sociology.
Wheale P and McNally R (1990) 'The consequences of modern genetic engineering: patents, "nomads" and the "bio-industrial complex"' in P Wheale, R von Schomberg and P Glasner (Eds.) *The Social Management of Genetic Engineering* Aldershot, Ashgate.
WIHSC (1997) *Citizens' Jury on Genetic Testing for Common Disorders: Recommendations* Glamorgan, Welsh Institute for Health and Social Care.

Wood-Harper J and Harris J (1996) 'Ethics of human genome analysis: some virtues and vices' in Marteau T and Richards M (Eds.) *The Troubled Helix. Social and Psychological Implications of the New Genetics* Cambridge, Cambridge University Press.

Wright S (1994) *Molecular Politics: Developing American and British Regulatory Policies for Genetic Engineering, 1972–1982* Chicago, University of Chicago Press.

Wynne B (1989) 'Establishing Rules of Laws: constructing expert authority' in Smith R and Wynne B (Eds.) *Expert Evidence. Interpreting Science in the Law* London, Routledge.

Wynne B (1991) 'Knowledges in Context' *Science, Technology and Human Values* 16, 111–121.

Wynne B (1995) 'Public understanding of Science' in Jasanoff S et al (Eds.) *Handbook of Science and Technology Studies* London, Sage.

Wynne B (1995a) 'Technology Assessment and Reflexive Social Learning: Observations from the risk field' in *Managing Technology in Society: the Approach of Constructive Technology Assessment* Rip A, Misa TJ and Schot J (Eds.), London, Pinter.

Wynne B (1996) 'May the Sheep Safely Graze? A Reflexive View of the Expert-Lay Knowledge Divide' in Lash S, Szerszynski B and Wynne B (Eds.) *Risk, Environment and Modernity. Towards a New Ecology* London, Sage.

Wynne B (1996a) 'Patronising Joe Public' *The Times Higher Educational Supplement* 12 April, 13.

Young IM (1990) *Justice and the Politics of Difference* Princeton N.J., Princeton University Press.

Ziman J (1984) *An Introduction to Science Studies* Cambridge, Cambridge University Press.

Zimmern R (1998) 'Genetic Medicine, Common Diseases and Public Health' in *New Paradigm: New Policies. Healthcare and the New Genetics in Britain and Germany* London, Genetics Interest Group, 4.

Zweiger G (2001) *Transducing the Genome: Information, Anarchy and Revolution in the Biomedical Science* New York, McGraw Hill.

Index

Abinex 101
Adam, B 86
adverse drug reactions (ADRs) 84
Advisory Commission on Human Genetics 114
Advisory Committee on Releases to the Environment 82
Affymetrix 23, 97, 104
  business model 105–6
Agilent Technologies 98, 106
Agricultural and Environment Biotechnology Commission 68–9
agriculture, and biotechnology 84–5
Alder Hey Children's Hospital 55
Alexion Pharmaceuticals 101
Alta meeting 13
Amersham Biosciences 24, 98
AmGen 96
animal growth hormones, and risk assessment 70–1
anonymisation, and genetic information 56
Applera Corporation 99
Applied Biosystems 22, 23–4, 33, 37, 99
Arabidopsis thaliana, sequence completed 35
Asgrow 82
Association of British Insurers 57
Aventis 104

Bauman, Z 78, 87, 88
Bayer 101, 104
Bayh-Dole Act (1980) 93
Beck, Ulrich 65, 86, 119
Beckett, Margaret 76
Beckman Coulter 106
Berg, Paul 23
'Bermuda Principles' 41
Bernal, J D 90
big science
  and Human Genome Project 9–11, 15–16, 19
  and megascience 16
  requirements of 123
biobanking 53–6
biodiversity 80–1
Biogen 12, 96, 101
bioinformatics 60–2, 107
BioMerieux Vitek 106
biotechnology
  business models of 95–6
  collective nature of 123–4
  and commercialisation 83–5, 124
  concentration of industry 78–9, 81
  concerns over 2–3
  and globalisation 79–81
  origins of 2
  paradigm shift 123, 125–6
  and pharmaceutical industry 94–5
  and public research 92
  and speculative investment 47
Biotechnology Science Co-ordinating Committee 16
Birmingham University 113
Blair, Tony 47, 48, 122, 126
Bloecker, Helmut 44
Bodmer, Walter 8, 15, 122
Bork, P 61–2
Botstein, David 19, 25
Bowker, G C 62
Boyer, Herbert 22, 93
Branscomb, Elbert 44
Brenner, Sydney 122
  and cDNA sequencing approach 29
  and Human Genome Project 9
  on technique 19
Bristol Royal Infirmary 55
British Biotechnology 96
British Medical Association 71
Brown, Patrick 23
Brownlea, A 111
BSE 66
Bud, Robert 2
business models, *see* genomics, business models

C Elegans, sequence completed 34
Caltech 22, 23–4
Cambridge, and 'Genome Fen' 109, 125
Cambridge Discovery Chemistry 105
Cantor, Charles
  and bioinformatics 61
  and Human Genome Project 15
Castells, M 78
Celera Diagnostics 99
Celera Genomics 13, 33, 34, 97, 99–100, 121
  challenges Human Genome Public Consortium 36–7
  commercial approach of 41
  criticisms of 35–6
  draft sequence completed 34–5, 48
  draft sequence published 35, 49–50
  formation of 33
  and National Institutes of Health 47–8
  as rival sequencing network 35
  seeks patents 41
  and shotgun sequencing 37
Cetus Corp 22
Chiron 96
Citizen Foresight Project 75, 113
citizens' juries 113, 119–20
  see also Welsh Citizens' Jury
citizenship
  and globalisation 87–8
  and the nation-state 77
  and rights 77
  see also citizens' juries
civil liberties 60
Clinton, Bill 3, 47, 48, 84, 122
CODIS 59
Cohen, Stanley 22, 93
Cold Spring Harbour Symposium 13
collaboratory, the 61
Collins, Francis 37, 45, 48, 126–7
  on Celera 38
  and Human Genome Project 30
  and priority disputes 46
  and Wellcome Trust funding 43
commercialisation
  and biotechnology 79, 83–5
  and genetic databases 54, 55, 57
commodification
  and genomes 41
  and genomics business models 95–6
  and knowledge 84, 122
  and patents 92–4
  and science 89–92
communication technologies, and globalisation 77
consensus conferences 115, 116
consent
  and DNA profiling 59
  and genetic information 55, 56, 57
  and genetic tests 57
controversies
  and revelation of assumptions 73
  and risk assessment 70, 72
  and science 67
  stages of 69
Cook-Deegan, R M
  and bioinformatics 60, 61
  and genomics firms 95, 107
  and origins of Human Genome Project 9, 12, 14
Coote, A 113, 119
COR Therapeutics 101, 105
Creutzfeld-Jakob Disease, new variant 66
Crick, Francis 2, 31
Criminal Justice and Police Act (2001) 60
CuraGen 101–2

Data Protection Act (1998) 60
Davis, Joel 12
Dawkins, Richard 7
De Waele, D 85–6
decision-making
  and citizens' juries 113–20
  public involvement in 74–6, 111, 112–13, 120, 125
deCode Genetics 54, 55, 57
DeLisi, Charles, and origins of Human Genome Project 13–14
Department of Energy (USA)
  and Human Genome Project 13–15, 34
  reorganisation of sequencing work 43–4
  and sequencer development 24
developing countries
  and genomics 108
  and global inequality 80
Dexter, Michael 47, 48
disability, and citizens' juries 119
discrimination, and genetic information 58–9
DNA profiling 59–60
Domenici, Peter 14
DoubleTwist 107
Dovichi, Norman 22
Drosophila melanogaster
  sequence completed 34
  sequencing project 38

Dulbecco, Renato, and Human Genome Project 13, 15
DuPont 59, 82, 102

Earthscan 88
Economic and Social Research Council 66
employment, and genetic information 58–9
English Nature 68, 69
Enzo Diagnostics 106
Epoch Biosciences 98
Equal Opportunities Act (1976) 57
Estonia, and genetic database 54
ethics
  and genetic information 56
  and globalisation 87–8
ethnicity, and citizens' juries 119
Etzkowitz, Henry 91, 92
eugenics 4
Eurobarometer 120
European Commission 70, 93–4
European Patent Office 80, 93
European Union, as super nation-state 78
experiments, and instrumentalities 18
experts
  communication repertoires 72–4, 116
  credibility of scientific 72
  and lay people 115
  and risk 66, 67
  and 'the tragedy of expertise' 72–4

family history, and genetic databases 55
Fixdal, J 115
Fodor, Stephen 105
food, *see* genetically modified food
Food Advisory Committee 71
Food and Drug Administration 81
France, and Human Genome Project 17
Franklin, Rosalind 2, 31
Friends of the Earth 68
Fujimura, J 62
Fukuyama, Francis 4

Galapagos Genomics 98
Gardner, David 11
Gemini Genomics 109
gender, and citizens' juries 119
Genentech 96, 102
Genethon 19, 23
Genetic Alliance 58
Genetic Analysis Technology Consortium 106
genetic engineering, origins of 2
Genetic Manipulation Advisory Group 70
Genetic MicroSystems 106
genetically modified food
  controversy over 65, 67–9
  crop trials 68–9
  and Food and Drug Administration 81
  and food labelling 71
  public involvement in decision-making 75–6
  and scientific expertise 72–4
genetically modified organisms, and risk assessment 82
genetics
  classical 1–2
  and eugenics 4
  see also genomics; information, genetic; new genetics
Genetics Forum 75
geno-technology
  implications of 1, 4
  and regulation 4
Genome Corp 13
genomics 3
  business models 95–6, 109
    enabling genomic technology 105–7
    functional genomics 100–2
    structural genomics 97–100
    target drug discover 102–5
  characteristics of firms 107–9
  and commodification 41, 90–1
  definition 89
  funding 107
  and new paradigm 123–6
  patents 108
  and pharmaceutical industry 95
  and speculative investment 47, 108–9
  and Third World 108
Genset 107
Genzyme 96
Gibbons, M 92, 125
Gibbs, Richard 43
Giddens, Anthony 85
Gilbert, G N 72, 74, 116
Gilbert, Walter 7
  and origins of Human Genome Project 11, 12–13
  and sequencer development 22, 23
GlaxoSmithKline 102, 103
GlaxoWellcome 81
globalisation
  and citizenship 87–8
  the global and the local 85–6

impact of 77, 87
and the nation-state 78
and polarisation 79–81
Goldberg, R 84–5
Gore, Al 48
Green, Phillip 23, 32
Greenpeace 68
Gulf War Syndrome 67

Haemophilus influenzae, sequencing of 31, 32
Haldane, J B S 1, 3, 4
Harrison, S 114, 120
Haseltine, William 41, 103
hazards, natural 65
Health and Social Care Act (2001) 56
Health Sector Database 54, 55
Healy, Bernadine 93
herbicides, and risk to farm workers 66
Hilgartner, Stephen 18, 25–6
Hill, A 114, 120
Hine, C 61
Hinxton 125
Hoffman-LaRoche 54, 102
Hood, Leroy 11, 22, 23
House of Lords, and policy-making process 112
Huang, Henry 24
Human Genetics Commission 55–6, 57
Human Genome Diversity Project 4
Human Genome Organisation (HUGO) 78–9
Human Genome Project (HGP) 121
applications of 122
as big science 9–11, 15–16, 19
'black boxing' of 122–3
as boundary object 122
and commercial pressures 42
development of 33–6
Celera's challenge 36–7
disagreements over strategy 29–30
draft sequence completed 48–50
draft sequence published 49–50
and future of 51
network reorganisation 43–5
public versus private 40–3
scientific race 45–8
sequence completed 50
sequencing methods 31–3, 37–9
funding of 39–41
and industrialisation of research 19
and instrumentalities 7, 18, 21–3, 37–9
and the Manhattan Project 4, 13
origins of 3, 27
DeLisi's role 13–14
and Department of Energy 13–15
doubts about 16
funding 17
Gilbert's role 12–13
goals of 17
and metaphors for 7–9
and National Institutes of Health 14–15, 16–17
Sinsheimer's role 9–12
and sequencer development 23–5
technological dependence 21
as technology 19–20
Human Genome Public Consortium 32
and 'Bermuda Principles' 41
and Celera's challenge 35–7, 43–5
composition of 36
draft sequence completed 34–5, 48, 50
draft sequence published 35
funding 43
and G5 group 43, 44
and media relations 47
regional approach 44–5
reorganisation of sequencing work 43–5
and shotgun sequencing 38
strategy 43
timetable 45
Human Genome Sciences 31, 41, 42, 97, 110
business model 102–4
Hunkapillar, Michael, and Celera 33

Iceland
and Health Sector Database 54, 55
and regulation of genetic databases 55
Iconix Pharmaceuticals 98
Incyte Genomics 97–9
Incyte Pharmaceuticals 42
inequality, and globalisation 80
information, genetic
and biobanking 53–6
and bioinformatics 60–2
commercial use 54, 55, 57
and DNA profiling 59–60
and employment 58
and insurance 57–8
and policing 59–60
and privacy 56–7, 60
and regulation 55–6
types of 53
Ingenuity 104
innovation

and genomics firms 109
models of 90–1
Institute for Genomic Research (TIGR) 31, 103
and shotgun sequencing 31–2
Institute for Public Policy Research 113
instrumentalities
access to 35, 37–9
definition 7
and Human Genome Project 7, 18, 21–3
role of 18
and science 17–18, 20–1, 50
and sequencer development 23–5
and social control 25–6
insurance, and genetic information 57–8
intellectual property rights 42, 80
see also patents
International Human Genome Sequencing Consortium, *see* Human Genome Public Consortium
Irwin, A 120

Jaklevic, Joseph, and sequencer development 24–5
Japan, and Human Genome Project 17
Joss, S 111
juries, *see* citizens' juries

Kanehisa, M 61–2
Karger, Barry 22
Kerr, A 115
King's Fund 113
knowledge
commodification of 84, 122
contingent nature of production 76
deficit model of 74–5
interdisciplinary approach 125
socially robust 111–12, 125
and standard of living 79

labelling, and food products 71
Lander, Eric 44
Lash, S, and risk 66–7
Lawrence Livermore national laboratory 14
Lederberg, J, and the collaboratory 61
Lee, Thomas F, and Human Genome Project 15
legitimation, and user involvement 120
Lehrach, Hans 44
Lenaghan, J 113, 119
Leukosite 105
Lewontin, Richard 123
Lexicon Genetics 99
Lindee, S 121
Linden Bioscience 99
LION Biosciences 99
London Centre for Governance Innovation and Science 75
Los Alamos national laboratory 14

McIver, S 118
McKusick, Victor 13–14
Macnaghten, P 85, 87
'Mad Cow Disease' 66
malathion 67
Manchester, University of, and UK Biobank 54
Manhattan Project 3–4, 13
mapping 11–12, 122
Marshall, T H 77
Marx, Karl 90, 91
Maxam, Alan, and sequencer development 22, 23
Medarex 99
media
and Public Consortium 47
reaction to Celera 40
and scientific communication 36–7
Medical Research Council, and Human Genome Project 19
megascience 51, 16
Meldrum, Peter 110
Mendel, Gregor 1
Merck Gene Index 42
Merck Pharmaceuticals 42, 80–1, 103
Merton, Robert 46, 74–5, 89
Michael, M 114, 120
Millennium Pharmaceuticals 97, 101, 104–5, 106, 110
Mitsubishi 102
MJ Research 23–4
Molecular Dynamics 23, 24
Monsanto 69, 72, 82, 102
Morgan, Michael 42
Morgan, T H 11
Mort, M 114, 120
Mouse Genome Consortium 106
Mulkay, M 72, 74, 116
Mullis, Kary 20, 22
Myers, Gene 32, 33
Myriad 110

NASDAQ 54
nation-state

and citizenship 77
and economic development 77–8
and genetic technologies 81
and globalisation 78
regulatory role of 81–3
role of 78
National Biodiversity Institute 80–1
National Centre for Genome Research 15
National DNA database 59–60
National Farmers Union 71
National Health Service 53, 126
National Human Genome Research Institute 126
National Institute for Clinical Excellence 83
National Institutes of Health (NIH)
and Celera 47–8
and expressed sequence tags approach 30
and Human Genome Project 8, 13, 34
and origins of Human Genome Project 14–15, 16–17
and patent applications 93
reorganisation of sequencing work 43–4
National Science Foundation (NSF), and Human Genome Project 16
Nature, and Human Genome Public Consortium 49–50
nature
boundaries of 1
distance from 85–6
and impact of globalisation 87
neem tree, and patents 80
Nelkin, Dorothy 74, 119, 121
Neomorphic 106
new genetics
commercial implications 2
data richness of 124–5
and the future 126–8
and geno-technology 1
nature of 121
origins of 1–2
threat of 119
see also genomics; information, genetic
non-governmental organisations 68
Novartis 72
Nowotny, H 111–12
Nuffield Council on Bioethics 68, 69, 94

Odyssey Pharmaceuticals 99
Office of Human Genome Research 15, 15–17
Office of Technology Assessment (OTA) 16
Ogilvie, Bridget 42
Olsen, Maynard 44
Olson, M 25
Ono Pharmaceuticals 102
Oppenheimer, Robert 3
Orchid BioSciences 106
organic food 85–6
organophosphates 66, 67
'Our Inheritance, Our Future' 126

Paabo, S 123
Panagea Systems 107
Parliamentary Office of Science and Technology 112
patents 109
and biodiversity 80
and biotechnology 92–3
and Celera 42
and genomics 108
and human DNA 93–4
requirements for 92
and single nucleotide polymorphism 83
and universities 92–3
Patrinos, Ari 37
Pattersson, Mark 47
Pauling, Linus 46
PE Biosystems 23
PE Corporation, and Celera 33
PerkinElmer Life Sciences 99
Persidis, A 84
Pesticides Advisory Committee 66
Pfizer 81, 102
pharmaceutical industry 94–5
pharmacogenomics 83
pharmogenetics 83–5
policing, and genetic databases 59–60
Price, Derek de Solla, and instrumentalities 17–18, 123
privacy, and genetic databases 54, 56–7, 60
privatisation, and sequenced genome 42, 43, 89
proteomics 110, 125
Purdue, D 115, 116
Pusztai, Arpad 68, 69, 73, 116

Race Relations Act (1976) 57
racism, and genetic information 59
regulation
and genetic databases 55
and the nation-state 81–3
Relenza 83
representativeness, and citizens' juries 118–20

research
  changing modes of 91–2
  creation of new system 125
  industrialisation of 19
  management of 50–1
  politics of 30, 32–3
  public versus private 40–3
Richards, M, and family histories 55
Rifkin, Jeremy 4, 79, 81–2
rights, and citizenship 77
risk
  alternative views of 70–1
  changed nature of 65–6
  characteristics of 66–7
  and controversies 70, 72
  culture of 87
  and experts 66, 67
  and food production 65
  and GMO release 82
  perceptions of 66
  routinisation of assessment 70
  and science 66, 70
Robinson, Maxine, and polemics 42
Roche 104
Roche Molecular Systems 106
Roche Vitamins 102
Roderick, Thomas H 89
Roe, Bruce 47
Roos, D S 124
Rose, Hilary, and eugenics 4
Rose, N 120
Rosenberg, A 8
Rosenthal, Andre 44
Rothman, Barbara Katz 124
Rothman, H 121
Rowett Research Institute 68, 73
Royal Society 68, 73, 74, 111
Rubin, Gerry, and Celera 38–9

Sagar, A 80
Sainsbury, Lord 48
Sakaki, Yoshiyuki 44
San Francisco 125
Sanger Centre 30, 31, 34, 43
Sanger, Fred, and sequencer development 22, 23
Sanofi-Synthelabo 103
Santa Cruz, University of California, and origins of Human Genome Project 9, 10–11
Schering-Plough 96, 103
Schindler, Melvin, technology and scientific advance 20
Science, and Celera 49
science
  and commercial world 2, 41, 42, 51
  and commodification 89–92
  and communication 36–7
  and communication repertoires 72–4, 116
  contextual nature of 112, 116
  and controversy 67
  deficit model of knowledge 74–5
  erosion of traditional academic norms 42, 89
  and instrumentalities 17–18, 20–1, 50
  and interaction with technology 17–18
  and peer reviewed publication 49
  public involvement in decision-making 74–6
  public questioning of 111
  'Public Understanding of Science' 66
  races in 45–6
  and risk 66, 70
  social control processes 25–6
  and socially robust knowledge 111–12, 125
  and technology 20–1
  and 'the tragedy of expertise' 72–4
Science Museum 75
sequence-tagged sites (STSs) 25–6
sequencing 122
  clone-by-clone method 32
  controversy over methods 37–9
  and expressed sequence tags approach 29–30
  sequencer development 23–5
  shotgun 31–2, 37, 38
Sequenom 102, 109
Sergeant, Terri 58
Shapin, Steve 91
Shimizu, Nobuyoshi 44
Shiva, Vandana 81
sickle-cell anaemia, screening for 59
Silver, Lee 4
single nucleotide polymorphism (SNP) 83
Sinsheimer, Robert L 9–11
Smith, David 44
Smith, Hamilton 22
  and Craig Venter 31
  and shotgun sequencing 31–2
Smith, Lloyd 22
SmithKlineBeecham 81, 103
'snip-chip' 84

SNP Consortium 34, 43
social control
  and instrumentalities 25–6
  and surveys 120
soya, and genetically modified food 71
standardisation, and bioinformatics 62
Star, S L 62
state, *see* nation-state
Steinberg, Wallace 103
Stock, G 4
Sturtevant, A H 11
Sulston, John 34, 43, 47, 89
  on announcement of draft sequence 48
  on Celera 48
  and commercial pressures 42
  and expressed sequence tags strategy 29, 30
  on G5 group 44
  and mapping 12
  motivations of 45
  and origins of Human Genome Project 11
  and Sanger Centre 31
  and scientific competitiveness 46
  and sequencing 24, 38
  on Venter 39, 42
Sun Microsystems 107
'surveillance creep' 59
Synthelabo 103

Takeda Chemicals 103
Tate, W T 95
technology
  and closure 122
  and Human Genome Project 19–20
  and interaction with science 17–18
  public questioning of 111
  and scientific research 20–1
'Terri Sergeant Case' 58
Third World
  and genomics 108
  and global inequality 80
time, control of 86
Titmus, R M, and blood donation 57
trust, breakdown of 66

Uberbacher, E 23
UK Biobank 54
  and use of data 57
Uncapher, K, and the collaboratory 61
underclass, genetic 58
United Kingdom
  and biotechnology 79
  funding for genome project 40
  and genetically modified food controversy 67–9
  and Human Genome Project 17
  insurance and genetic information 57–8
  and 'Mad Cow Disease' 66
  and National DNA database 59–60
  'Our Inheritance, Our Future' 126
  and regulation of genetic databases 55–6
  and risk assessment 82
  and UK Biobank 54, 57
United Nations
  Convention on Biological Diversity 80
  Declaration on the Rights of Indigenous Peoples 88
United States
  and biotechnology 79
  funding for genome project 39
  genetic information
    and employment 58–9
    and insurance 58
  and genetically modified products 69, 79, 81
  and genomics firms 108
  law and regulation 82
  and origins of Human Genome Project 9–17
  and patent law 92
universities
  and biotechnology start-ups 91
  and life science centres 125
  and patents 92–3
Urry, J 85, 87

Variagenetics 83
Varmus, Harold 93
Venter, Craig 48, 93
  and Celera 33, 35, 99, 100
  challenges Human Genome Public Consortium 36–7
  criticisms of 42
  and expressed sequence tags strategy 29–30
  and Hamilton Smith 31
  seeks private funding 31

Wade, Nicholas 37
WarnerLambert 81
Waterston, Bob 30, 34, 35, 43
Watson, James 2, 46, 93
  and Francis Crick 31

heads Office of Human Genome Research 16–17
and Human Genome Project 12, 15
leaves National Institutes of Health 30
on Venter 42
and Wellcome Trust funding 43
Weber, James 32, 33
Webster, A 127, 128
Weissenbach, Jean 44
Weizmann, Chaim 2
Welfare State 77
Wellcome Trust 30, 34, 42, 48
and funding for genome project 40
and Sanger Centre 43
and SNP Consortium 43
Welsh Citizens' Jury 114
and communications repertoires 116
procedures 117–18
and pseudo-expertise 116
and representativeness 118–19
role of local knowledge 114–15
symbolic value 116–17
terms of engagement 115
Welsh Institute for Health and Social Care 114
White, Tony 37
Whitehead Institute 44, 106
Wilcox, Kent 22
Wilkins, Maurice 2, 4, 31
World Trade Organisation (WTO) 79
Wyngaarden, James 14
appoints Watson 16
on Department of Energy 14
and origins of Human Genome Project 15
Wynne, B 66–7, 74

Zeneca 72
Zimmern, R 125
Zweiger, G 124